Rad-hard Analog ICs

RIVER PUBLISHERS SERIES IN ELECTRONIC MATERIALS, CIRCUITS AND DEVICES

The "River Publishers Series in Electronic Materials, Circuits and Devices" is a series of comprehensive academic and professional books which focus on theory and applications of advanced electronic materials, circuits and devices. This includes analog and digital integrated circuits, memory technologies, system-on-chip and processor design. Also theory and modeling of devices, performance and reliability of electron and ion integrated circuit devices and interconnects, insulators, metals, organic materials, micro-plasmas, semiconductors, quantum-effect structures, vacuum devices, and emerging materials. The series also includes books on electronic design automation and design methodology, as well as computer aided design tools.

Books published in the series include research monographs, edited volumes, handbooks and textbooks. The books provide professionals, researchers, educators, and advanced students in the field with an invaluable insight into the latest research and developments.

Topics covered in the series include:-

- Analog Integrated Circuits
- Data Converters
- Digital Integrated Circuits
- Electronic Design Automation
- Insulators
- Integrated circuit devices
- Interconnects
- Memory Design
- MEMS
- Nanoelectronics
- Organic materials
- Power ICs
- Processor Architectures
- Quantum-effect structures
- Semiconductors
- Sensors and actuators
- System-on-Chip
- Vacuum devices

For a list of other books in this series, visit www.riverpublishers.com

Rad-hard Analog ICs

Umberto Gatti
RedCat Devices, Italy

Cristiano Calligaro
RedCat Devices, Italy

River Publishers

Routledge
Taylor & Francis Group
NEW YORK AND LONDON

Published 2026 by River Publishers
River Publishers
Broagervej 10, 9260 Gistrup, Denmark
www.riverpublishers.com

Distributed exclusively by Routledge
605 Third Avenue, New York, NY 10017, USA
4 Park Square, Milton Park, Abingdon, Oxon OX14 4RN

Rad-hard Analog ICs / Umberto Gatti and Cristiano Calligaro.

Routledge is an imprint of the Taylor & Francis Group, an informa business

ISBN 978-87-702-2419-2 (hardback)
ISBN 978-87-438-1332-3 (paperback)
ISBN 978-87-438-1329-3 (online)
ISBN 978-87-438-1331-6 (e-pub)
ISBN 978-8-743-81328-6 (ebook master)
DOI: 10.1201/9788743813286

Contents

Preface

Rad-hard Analog ICs is the natural prosecution of our previous book, Rad-hard Semiconductor Memories, and it is the result of more than a decade of efforts spent in designing and testing very different chips and embedded macros including ADCs, DACs, Bandgap References and other elementary blocks to be included in rad-hard mixed signal ASICs.

In this journey we learned that Radiation Hardening By Design has indeed a common origin, but mitigation techniques can be very different when moving from a digital scenario to an analog one and TID (Total Ionizing Dose) is the main threat.

In a digital block (a standard cell or a bistable circuit) the effect of charged traps can be "tolerated" because at the end what is provided is a logic "1" or a "0"; in an analog block the effect is much more subtle and insidious, and this is intrinsic in the nature of the continuous time and amplitude approach respect to the discrete amplitude of a digital world.

A differential stage with active load (the main block of a comparator or a differential amplifier) is prone to mismatches coming from the manufacturing process, the layout design, the operative environment (temperature, power supply) and other second order effects.

Analog designers are usually focused on making circuit topologies and physical layouts fine-tuned to compensate offsets (e.g. stacked transistors), but this is not enough when designing rad-hard analog ICs.

What we discovered is that process mismatches and charge traps are fused together, and process mitigation techniques may collide heavily with total dose mitigation techniques.

This is the arena where we had to fight: canonical solutions like stacked transistors and matched current mirrors simply can't be used because they are killed by intra-device and inter-device leakage effects; but on the other hand, effective hardening solutions like shielding or physical separation between N-channel and P-channel transistors open the road to detrimental process deviations that are intrinsic in any manufacturing process.

We have asked ourselves many times, in the typical Hamlet-like dilemma, if it was more acceptable to suffer the process mismatches or the total dose degradation not considering that also single event effects play their role.

Moreover, many architectural solutions adopted in ADCs and DACs can be very difficult to be protected in hostile environments and, as always happens, some old-fashion (let's say Vintage!) architectures appears to be more resilient with respect to the state of the art. This obviously comes at the expense of performance, but all engineers have in their DNA the ability to find the best trade-off.

We don't claim to have answered all these questions; in fact, many remain unanswered. But this is only the beginning of the journey.

Umberto Gatti and Cristiano Calligaro

Acknowledgments

Part of this book is the result of research activities in the framework of Italian Space Agency (ASI) funded project RELIC (CUP F43D24000030001 and CIG B05507A50B) for Chapters 1, 2, and 5 and in the framework of Space R&I Work Program of the European Commission (EC) 870365 MORAL Project for Chapters 3, 4, and 5.

The authors would like to thank ASI project officer Simone Pizzurro (for RELIC Project) and European Commission project officer Simon Silvio Conticello (for MORAL Project).

Testing under irradiations would not be possible without the support of Dr. Aldo Parlato and Prof. Elio Tomarchio from University of Palermo (Italy) for Cobalt 60 total dose tests and Heikki Kettunen, Mikko Rossi, and Jukka Jaatinen from University of Jyväskylä (Finland) for single-event heavy ions tests. The authors would like to thank them all.

Finally, the authors would like to express their sincere gratitude to Armando Colamonico and Marcello Donati for all the pre-irradiation in-lab characterization of the developed electronic components.

List of Figures

List of Tables

List of Notations and Abbreviations

ADC	Analog-to-digital converter
ADPLL	All digital phase-locked loop
ASET	Analog single-event transient
BGR	Bandgap reference
BiCMOS	Bipolar mos
BICS	Bulk built-in current sensors
BJT	Bipolar junction transistor
BW	Band width
CDS	Correlated double sampling
CERDIP	Ceramic dual-in-line package
CMOS	Complementary metal oxide semiconductor
CTAT	Complementary to absolute temperature
CTS	Clock tree synthesis
DAC	Digital-to-analog converter
DAT	Digitally assisted techniques
DB	Daughter-board
DBLC	Dynamic base leakage compensation
DBT	Dissociated binary tree
DD	Displacement damage
DEC	Digital error correction
DGA	Dummy gate assisted
DMR	Dual modular redundancy
DNL	Differential non-linearity
DTMOST	Dynamic threshold mos transistor
DUT	Device under test
DWA	Dynamic weighted averaging
EGR	Extended guard-ring
ELT	Edge-less transistor
ENOB	Effective number of bit
FFT	Fast fourier transform
FIR	Finite impulse response
FPGA	Field programmable gate array

GUI	Graphic user interface
HBT	Heterojunction bipolar transistor
HEP	Hole electron pair
IIR	Infinite impulse response
IM	Inverse-mode
IMD	Inter-modulation distortion
INL	Integral non-linearity
LDO	Low-dropout
LEO	Low-earth orbit
LET	Linear energy transfer
LOCOS	Local oxidation of silicon
LSB	Least significant bit
MB	Mother-board
MBU	Multiple bit upset
MCU	Multiple cell upset
MIM	Metal-insulator-metal
MOSFET	Metal oxide semiconductor field effect transistor
MSB	Most significant bit
NISO	N-well isolation
NMR	N-modular redundancy
NSB	Next-most significant bit
OSR	Over-sampling ratio
OTA	Operational transconductance amplifier
PLA	Programmable logic array
PSRR	Power supply rejection ratio
PTAT	Proportional to absolute temperature
PVT	Process, voltage, temperature
RA	Residue amplifier
RHBD	Radiation-hardened-by-design
RHBP	Radiation-hardened-by-process
RMS	Root mean square
ROM	Read only memory
SAR	Successive-approximation register
SC	Switched-capacitor
SCR	Silicon controlled rectifier
SEE	Single-event effect
SEFI	Single-event functional interrupt
SEL	Single-event latch-up
SET	Single-event transient
SEU	Single-event upset

SFDR	Spurious-free dynamic range
SINAD	Signal-to-noise and distortion
SMA	Subminiature version a
SNR	Signal-to-noise ratio
SOI	Silicon on insulator
SOS	Silicon on sapphire
SRAM	Static random access memory
STI	Shallow trench isolation
TDC	Time-to-digital converter
THA	Track-and-hold amplifier
THD	Total harmonic distortion
TI	Time interleaved
TID	Total ionizing dose
TMR	Triple-modular redundancy
UWT	Ultra-wide temperature
VPN	Virtual private network
ZIF	Zero-insertion force

1

Radiation Effects and RHBD Techniques

Abstract

In this chapter, an overview of threats coming from space and affecting semiconductor components with a specific focus on CMOS will be described. Hard errors and soft errors will be introduced as keywords of the overall description of radiation effects and mitigation efforts to obtain a rad-tolerant or a rad-hard component.

1.1 Energy Release from Invisible Enemies

It is quite common in the space community considering radiation effects on silicon components as something obscure, complex, and requiring fundamental skills in nuclear physics or similar disciplines. Maybe this is true for those willing to describe such effects in a theoretical way, but for semiconductor engineers willing to design components, only some fundamentals are required to pave another (and maybe more challenging) road represented by the mitigation of radiation effects on silicon.

Coulomb interaction and lattice displacement are the two main effects to be considered in this context.

Coulomb interaction is represented by energy release of charged particles (electrons, protons, ions, photons) colliding with silicon (whenever it is doped silicon or silicon dioxide). Indeed, the same effect exists also on Earth (for example, just exposing the chip to the Sun), but the difference is that in space, energy release is higher … much higher. This effect is not mechanical (between protons and electrons inside an atom, there is plenty of space), but the simple passage of such particles enables an energy transfer (via Coulomb field) to electrons and when this transfer is energetic enough, they can be extracted, thus creating the so-called hole-electron pair (HEP).

Figure 1.1 represents a generic atom with electrons orbiting and a charged particle track (it can be an alpha particle a photon or a heavy ion)

DOI: 10.1201/9788743813286-1

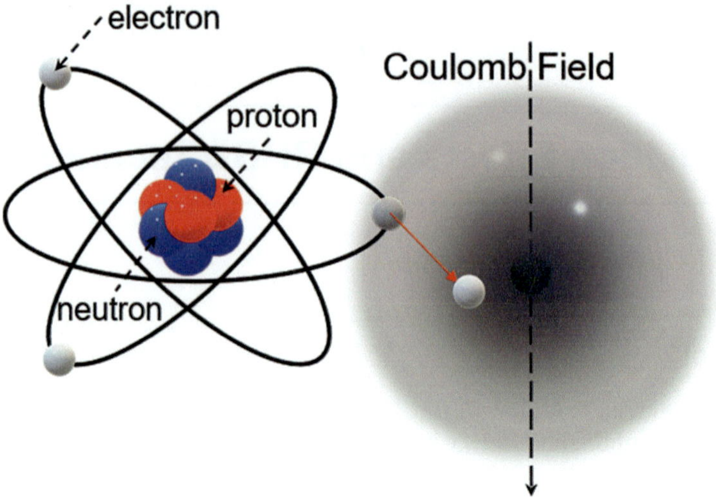

Figure 1.1 Hole-electron pair (HEP) generation.

passing nearby. The charged particle has a field (Coulomb field) influencing electrons moving along their orbitals; the higher is charge of the particle, the larger is the Coulomb field, and if the energy is high enough, it can extract the electrons from the orbital, thus leaving the atom without it: a pair is created until another electron can reach the vacancy and restore the previous stable state.

Holes and electrons as independent entities of course try to recombine, but it must be also underlined that their mobility is very different; electrons, thanks to their small size, can move in the lattice with more agility than their counterparts. This aspect is crucial when external factors intervene; one of these is the electric field, and in a CMOS device, this always happen when either in digital or analog operations.

In Figure 1.2 is shown, just like Figure 1.1, the passage of a charged particle in a generic media (e.g., silicon) and how its track separate electrons from the rest of the atoms, thus leaving behind several HEPs willing to move, in the beginning, in the direction of charged particle (or to be more precise in the direction of its Coulomb field) and, when the track is over, in the direction of any positive source they can find nearby.

Of course, if the media is not biased (no electric field), the closest positive source is the original atom they left (or other atoms in proximity), and in that case, they move forward to recombine and find the original stable state as shown in Figure 1.3.

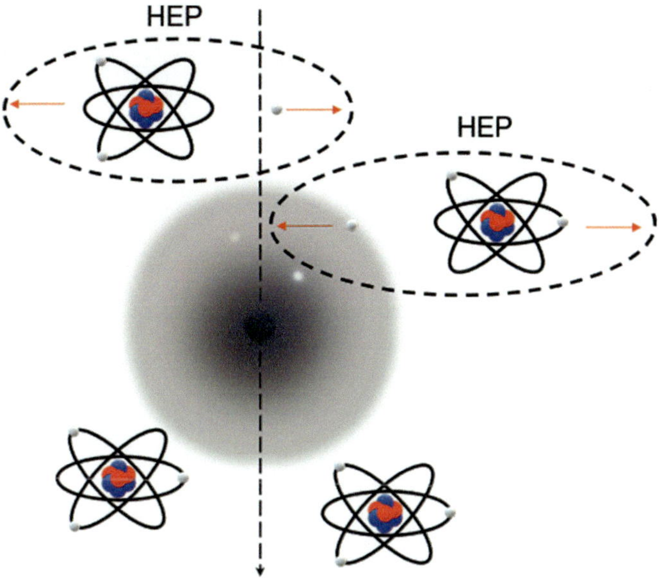

Figure 1.2 HEP generation.

Without electric field, recombination is fast and everything can be seen as the passage of a Coulomb field creating a sort of cloud of electrons and holes that recombine immediately after the field is over. Nothing can be seen outside, but when an electric field is applied (Figure 1.4), things change.

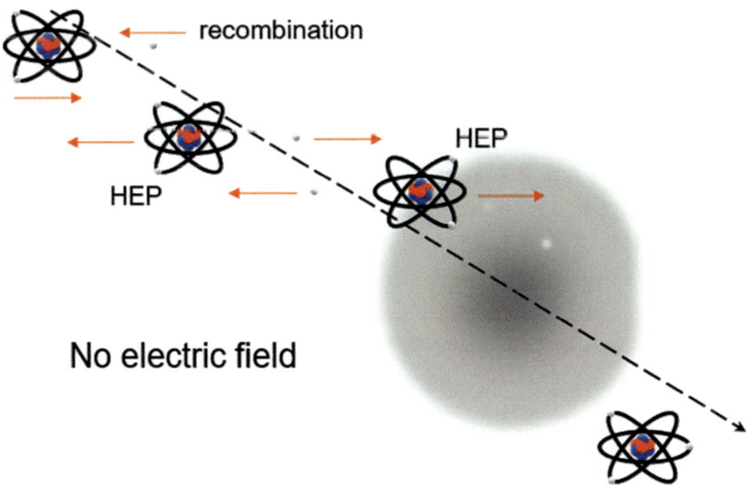

Figure 1.3 HEP generation and recombination.

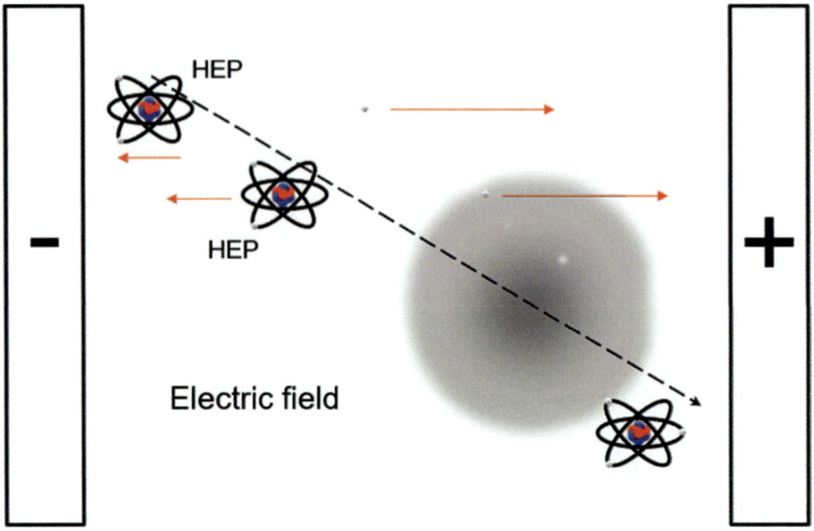

Figure 1.4 HEP generation in the presence of an electric field.

HEPs generated by the charge particle track, once the Coulomb field is gone, have now new attracting nodes, and what now intervenes is the distance and energy gained by the electron during the passage; if the energy transferred has been low, then recombination dominates, but if the energy is high, enough electrons start moving in the direction of the positive node and holes in the opposite direction. Of course, mobility is different and electrons may reach their target with higher probability. This is crucial in media like silicon dioxide (SiO_2) because for holes, it is much harder to move in an insulator than their negative counterparts, and if they found other atoms (e.g., defects in dioxide), they can establish a sort of positive island; this is called charge trap and it is the mechanism explaining the degradation of a silicon component along time or total dose ionization (TID). Charge traps are generated by HEPs where the positive counterpart (the hole) instead of finding an electron for recombination is attracted to impurities in the media, and this happens in proximity of the interface between silicon dioxide and doped silicon; months or years of exposure to particles releasing energy and creating HEPs progressively produce a collection of traps becoming biasing regions and affecting the behavior of transistors or other silicon-based active components (diodes, power MOSFET, non-volatile memory cells, etc.).

Oxide degradation coming from charge trapping effect (TID) is one of the most important parameters to be considered when designing or evaluating a rad-hard component to be used for space applications. Silicon dioxide,

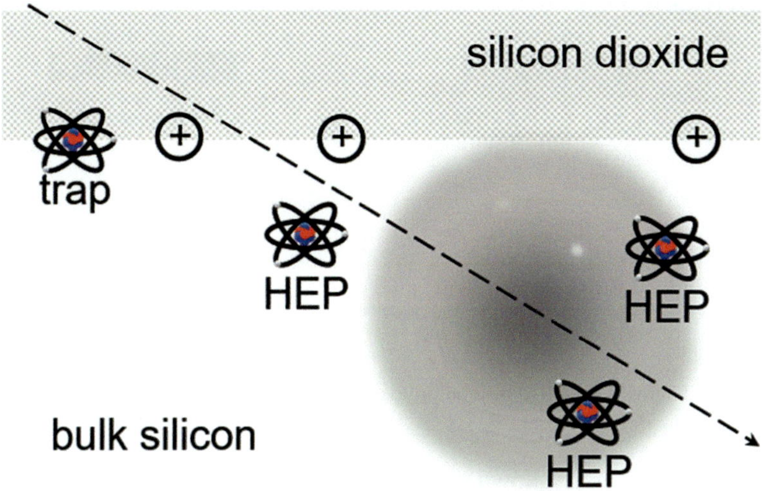

Figure 1.5 Charge trapping at bulk silicon and dioxide interface.

at wafer manufacturing level, is obtained in several ways; CVD (chemical vapor deposition) is one of the most common, but also thermal deposition in presence of water is used. Of course, these methods can produce hydrogen ions and these ions may become a defect producing a charge trap when HEPs are generated with the strike of a charged particle. Thermal deposition may be wet (most common) or dry (less common): the latter is less prone to hydrogen ion generation during manufacturing [2], thus strongly reducing the probability to have charge traps. Experiments in laboratory demonstrated that 1 Mrad (Si) can be achieved for TID, but of course in a standard CMOS commercial flow, this value can be lower and strongly dependent on the node.

Oxide deposition in process over 180 nm (e.g., 250 or 350 nm) is thick and the probability to have defects is much higher than in thinner oxides like those in 65 nm or below, and this aspect strongly affects the mitigation techniques to be adopted in the design of a rad-hard component.

TID degradation is the major concern in submicron processes (those down to 100 nm) while in deep submicron (from 100 nm down to 28 nm) are less affected by charge traps thanks to very thin (and clean) oxides. FinFET processes (16 nm down to 4 nm) and gate-all-around (GAA) below 4 nm suffer TID in a very limited way.

Figure 1.5 describes the behavior of HEPs generation in proximity of silicon dioxide (in the gate of the transistor or in isolations between transistor), but also in doped silicon, they can produce an effect.

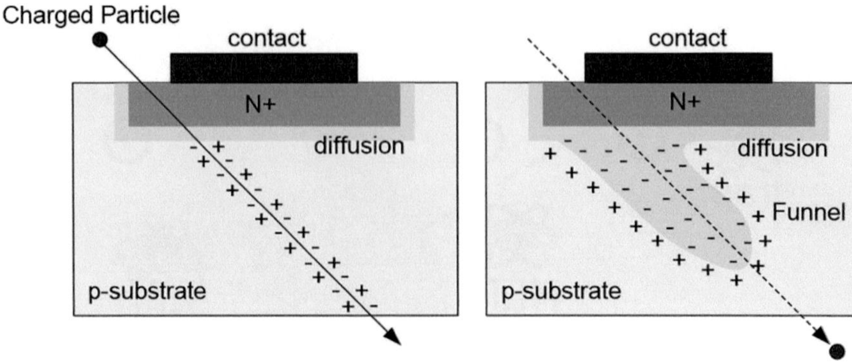

Figure 1.6 HEP generation in doped silicon and diffusion area.

In Figure 1.6 is shown a more complex structure (without oxides) including a P-doped substrate a strongly N-doped well (N+) and a diffusion area from N+ coming from a biasing. A contact completes a basic device very close to an NMOS transistor.

HEP generation has different effects according to the place where they are generated. In those parts where doped silicon creates drain and source regions, the charged particle track creates the "Funnel" region, a sort of cloud of electrons extending from the highly doped areas down to the bulk of the transistor (Figure 1.7). The same funneling effect can be observed from deep N-Well regions (if available in the CMOS process) into P-doped regions.

Funnel regions are the media through which currents are created. The energy release of the charged particle creates a drift current, while the natural recombination of holes and electrons creates a diffusion current. The two components are different in times (drift currents are generated in few

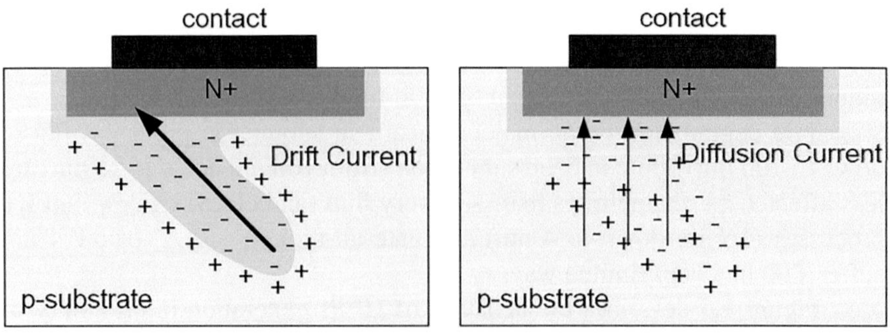

Figure 1.7 Drift and diffusion currents.

picoseconds while diffusion currents are over after a couple of nanoseconds), but the overall result is that energy release has been transformed in a current spike responsible of a charge injection.

HEPs in silicon dioxide have a different effect. If in doped silicon holes and electrons are recombined or migrated to positive/negative terminals in dioxides, HEPs have to consider mobility issues during migration when an electric field is applied. Electrons, thanks to their size, can move with a higher mobility than holes, and this takes to a natural recombination where electrons are the players, but the higher mobility can also distract them if an electric field is applied and this agility move them far from their half of the pair. In this case, holes still move in dioxide to find a recombination, but in their path, they can find defects of the lattice able to stop them and trap. As mentioned above, these are called charge traps and the responsibility of the effect is the manufacturing process of silicon dioxide (LOCOS or STI).

In general, the thicker is the oxide, the higher is the probability to find defects and of course the higher is the probability a hole is trapped. The thinner is the oxide, the lower is the probability to have defects. This is the main reason why deep submicron (<100 nm) technologies are less prone to charge traps (but on the other side more prone to transient as we will see in the next paragraphs).

It does not matter if they are generated in silicon dioxide or doped silicon: HEPs are always responsible of errors in semiconductor components that can be divided in two families:

a) Total ionizing dose (TID);

b) Single event effects (SEE).

The first are those errors coming from the exposure of semiconductor components to radiation sources [1], and the second includes errors coming from specific particles impinging silicon devices [2]. TID is a cumulative effect (the more prolonged is the exposure, the higher is the probability to degrade the chip) while SEE is stochastic.

Semiconductor components may have two different kinds of errors: (a) hard errors and (b) soft errors. Hard errors cannot be recovered (like latch-up), and soft errors are those giving a temporary error such as a bit flip or simply a transient propagating in the chip and that can be recovered via software or passing through hardware correcting systems (EDAC, error detection and correction). It is evident that radiation hardening by design (RHBD) techniques must, on top of all, guarantee a mitigation of hard errors and then a reduction of soft errors.

1.2 Cumulative Effect: TID

Standard CMOS transistors (both n-channel and p-channel) use silicon dioxide widely in the overall front-end of the chip. Silicon dioxide is used to isolate polysilicon gates from transistors' channels and to isolate contiguous transistors differently biased. In all cases, the key role is isolation and the better manufacturing the CMOS process is, the better is the behavior of the transistor itself and its independence from any biasing coming from neighbor transistors.

Silicon dioxide manufacturing is one of the key processes in a silicon foundry CMOS flow, and relevant efforts making pure oxides have been done in the past to obtain a low number of defects both in thin (channel) and thick (isolation) by using different techniques (e.g., dry oxidation).

In old manufacturing processes, isolation between transistors was guaranteed by local oxidation of silicon (LOCOS) where the passage from thin gate oxide to thick oxide was gradual and shaping the "bird's beak" [3] (Figure 1.8).

With modern technologies (not so modern nowadays!), a different technique, called shallow trench isolation (STI), overrode LOCOS, making more scaled and efficient isolation between contiguous transistors.

Charge trap phenomena are almost the same, both in gate oxides and isolation oxide, and whenever it is LOCOS or STI, the effect is the same.

In Figure 1.9 is sketched a typical n-channel transistor in a CMOS process using STI. Drain and source regions (collectively called active areas or active regions) are surrounded by silicon dioxide to guarantee isolation

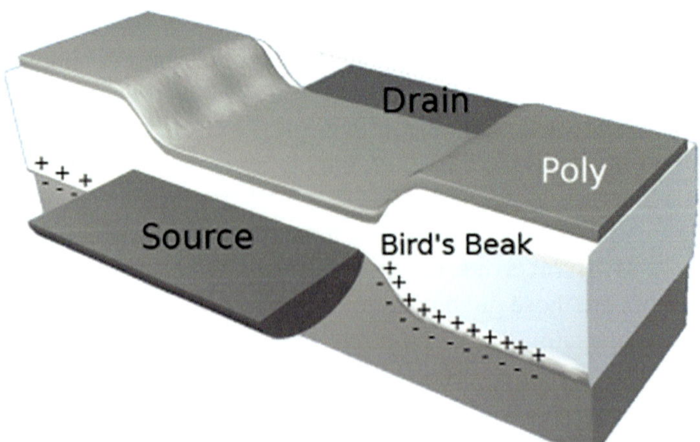

Figure 1.8 Local oxidation of silicon (LOCOS) and charge traps.

from other transistors biased at different voltages. Even if CMOS is a planar technology, nevertheless, STI thickness must guarantee the isolation blocking any biasing path passing through the bulk (p-substrate) of the chip. Thick oxide means a higher probability having defects and, consequently, giving rise to charge traps in case of holes generated by charged particles.

The same may occur in thin gate oxide (below polysilicon) responsible of the modulation of gate to create the channel path for the passage of charges from source to drain, but considering this time the oxidation is thin, a lower number of defects are expected, thus decreasing the probability to have charge traps. Deep submicron CMOS processes (below 100 nm) are considered by the Space Community safe enough to consider negligible any TID effect at transistor channel level in case LEO (low Earth orbit) applications are considered. Of course, the same cannot be said for STI considering the thickness they still have to guarantee isolation between neighbor transistors.

Since charge traps are responsible of leakage paths, we divide TID effects in two: (a) intra-device leakage and (b) inter-device leakage.

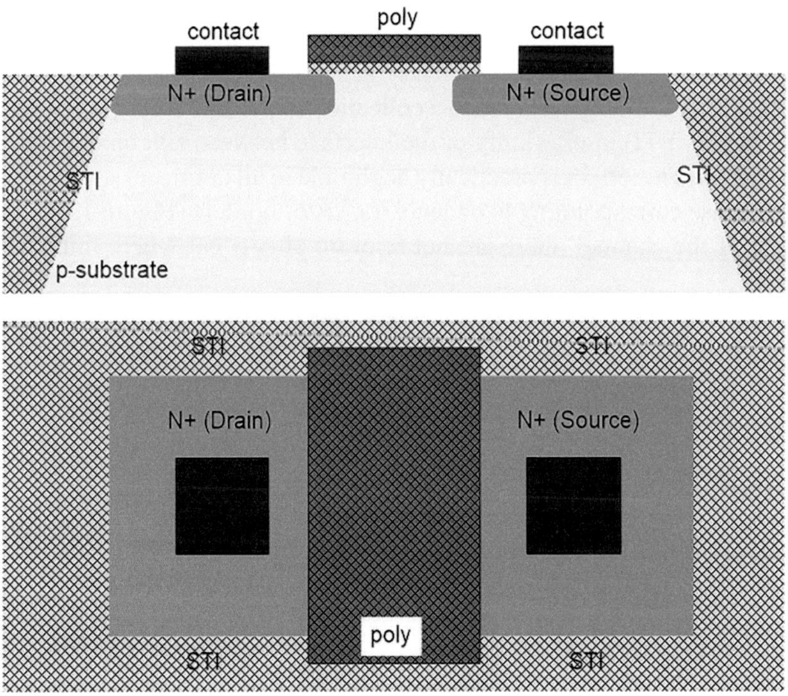

Figure 1.9 Cross section and top view of a standard n-channel transistor with STI.

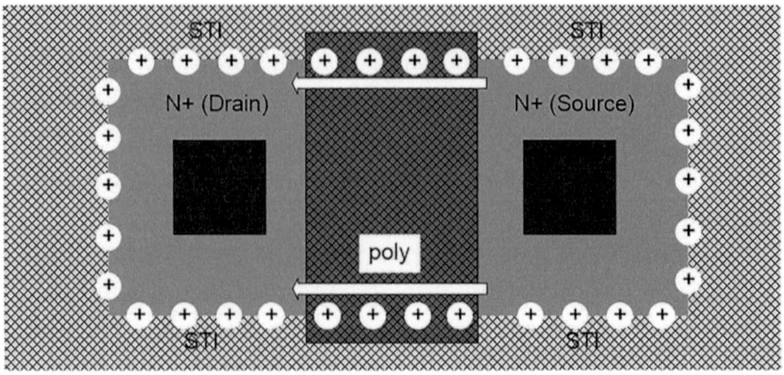

Figure 1.10 Intra-device leakage effect on transistor borders.

Intra-device leakage affects the single transistor only shaping its width and length and modifying its general performance; inter-device leakage affects different transistors for a biasing of the first affecting the performance of the second (and vice versa). Inter- and intra-device leakage may also combine their effects, leading to a further degradation that, it is worth underlining, is not recoverable (hard error).

In Figure 1.10 is sketched the effect of charge trapping in STI isolating a single n-channel transistor in a typical CMOS technology.

The progressive effects of HEP generation along time (month after month and year after year) lead to a collection of positive traps (holes trapped by defects in STI) in proximity of the interface between silicon dioxides and active areas. This effect is practically negligible in all drain and source regions except those corresponding to channel transistor borders (Figure 1.11). In the middle of the channel, there are not relevant effects but where thin channel

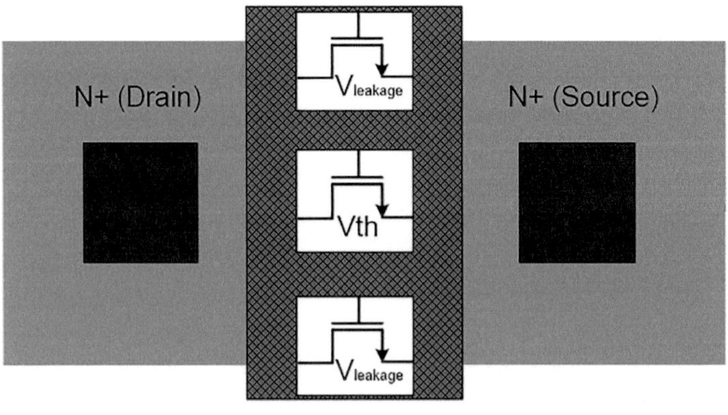

Figure 1.11 Threshold deviation in an n-channel transistor due to intra-device leakage.

oxide becomes thick isolation oxide (Bird's Beak or STI), a conductive path can bring charges from source to drain overriding the role of transistor gate.

From the electrical point of view, the overall effect is a deviation of the threshold voltage of the transistor. In the middle of the channel, the threshold is only partially or negligibly affected by charge traps, while in the borders, the effective threshold is decreased to a lower value; the overall threshold will be the medium value of the three contributes. Once the leakage paths on border will be strong enough to show $V_{leakage}$ below zero, the transistor will not be able to switch on and off its status: it has become a leaky transistor and its operations are compromised [1].

Concerning p-channel transistor, the situation is the same in terms of charge trapping but opposite in terms of threshold shift.

Along transistor channel borders, positive traps populate the region where thin oxide becomes thick oxide, but since this time along the channel are holes instead of electrons to move from drain to source what in the n-channel was a conductive path in the p-channel becomes a resistive path. In this case, the threshold of parasitic transistors on the borders will increase and current flow will become lower; when threshold voltage will override power, supply transistor will be always be in cut-off (Figure 1.12).

Figure 1.13 shows the threshold shift of an n-channel transistor working in pinch-off (or saturation) region before and after irradiation.

Inter-device leakage involves neighbor transistors and the effect is the biasing between different terminals. In Figure 1.14 is sketched an example of a pair of n-channel transistors separated (and isolated) by STI. In this

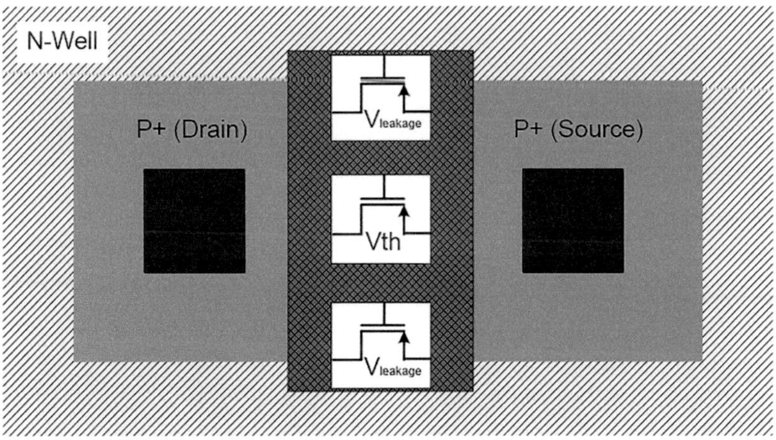

Figure 1.12 Threshold deviation in a p-channel transistor due to charge trapping.

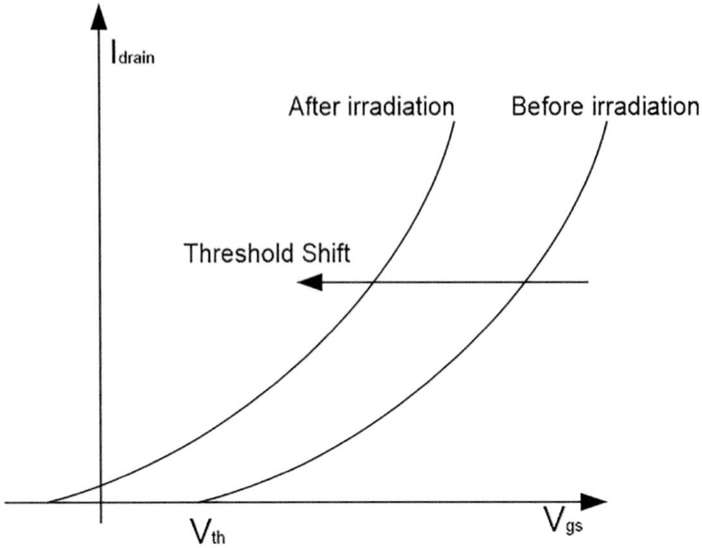

Figure 1.13 Threshold shift of an n-channel transistor in pinch-off region ($V_{ds} > V_{gs} - V_{th}$).

case, HEP generation coming from the cumulative effect of charged particles hitting the device creates charge traps that remain stable with their positive charge at the interface between silicon dioxide and p-substrate of the chip. The thickness of trench in modern CMOS technologies is enough to avoid any possible biasing in standard conditions thanks to the depth of the trench, but the growing number of traps at the interface creates a leakage path getting in communication N+ areas of different transistors. In case of a terminal having a specific voltage applied, the second may be affected. This phenomenon

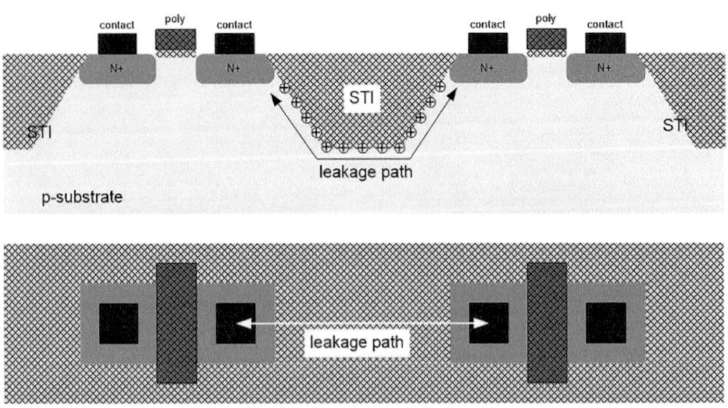

Figure 1.14 Inter-device leakage between neighbor n-channel transistors.

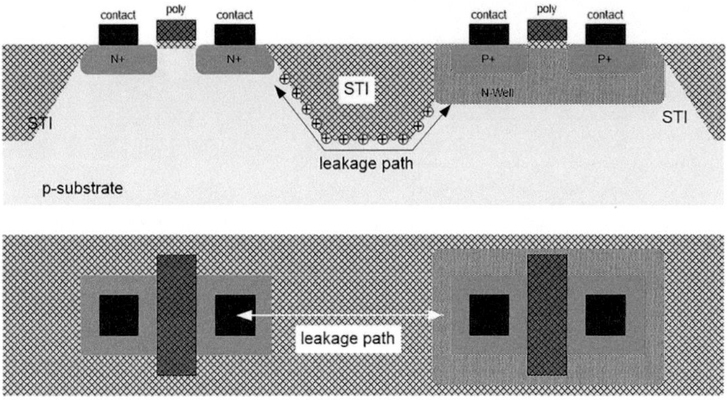

Figure 1.15 Inter-device leakage between n-channel and N-Well of a p-channel transistor.

is particularly dangerous in analog electronics where few millivolts can affect the working conditions of, for example, an operational amplifier or a comparator or a bandgap reference.

A similar situation can occur when an n-channel transistor is close to a p-channel transistor (Figure 1.15). In this case, the leakage path can make a connection between the N+ diffusion of one of the terminals of the n-channel transistor and the N-Well containing the p-channel transistor. In this case, if, for example, the N-Well is biased at power supply, the leakage path may bias the n-channel transistor terminal moving to a different voltage value.

1.3 Single Event Latch-up: SEL

In space components, SEL is considered the major threat, and many efforts are spent to make silicon devices resilient to this effect. The mechanism relies on vertical and lateral parasitic bipolar transistors coming from N-doped and P-doped regions.

In a standard CMOS process, parasitic bipolar transistors are the result of differently doped regions containing n-channel and p-channel transistors. P-channel transistors are those having drain and source heavily doped (P+), but since the substrate of the silicon wafer is lightly doped with the same dopant (P), it is necessary that they are included in a well-doped N: this region is called N-Well. On the other side, n-channel transistors, having drain and source heavily doped N (N+), can be built directly on the P-doped substrate of the silicon wafer.

This sequence of N-doped and P-doped regions can build bipolar transistors. On p-channel side starting from P+ diffusion of a drain/source and passing through the N-Well, the P-substrate is reached, thus giving a P-N-P junction

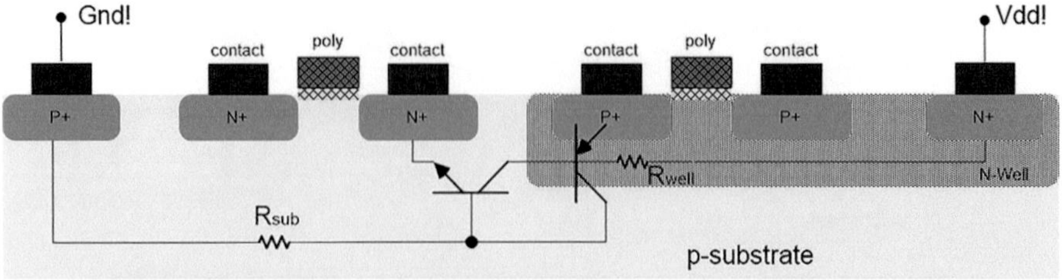

Figure 1.16 Silicon controlled rectifier (SCR) in a CMOS standard process.

also called vertical PNP bipolar transistor. On the n-channel side starting from N+ of a drain/source and passing through the substrate, it is possible to reach the N-Well: in this case, there is an N-P-N junction also called lateral NPN bipolar transistor. Moreover, this pair has two common nodes: the base of lateral npn connected to the collector of the vertical pnp (the p-substrate) and the collector of lateral NPN connected to the base of the vertical pnp (the N-Well). This connection shapes the typical positive feedback loop, and the structure is also called SCR (silicon controlled rectifier) or thyristor (see Figure 1.16).

SCR is normally in off condition and current flowing is negligible. This state is also called "blocking state," but it is possible to exit from this condition if a voltage is applied to the intermediate P terminal (node G in Figure 1.17): a sort of control gate. In this case, a current can flow from P-anode (node A) to N-cathode (node C) [2]. When triggered into the conductive state, the SCR enters the "latched state" and this comes from a current injection in the base of the NPN transistor (in Figure 1.16, this is the p-substrate) that, as a result, creates a current flow in the base-emitter junction of the PNP transistor, which turns on and increases further the current injected in the NPN transistor: the positive feedback loop is triggered and both transistors are saturated; the current flowing in both transistors guarantee that SCR does not exit from the latched state.

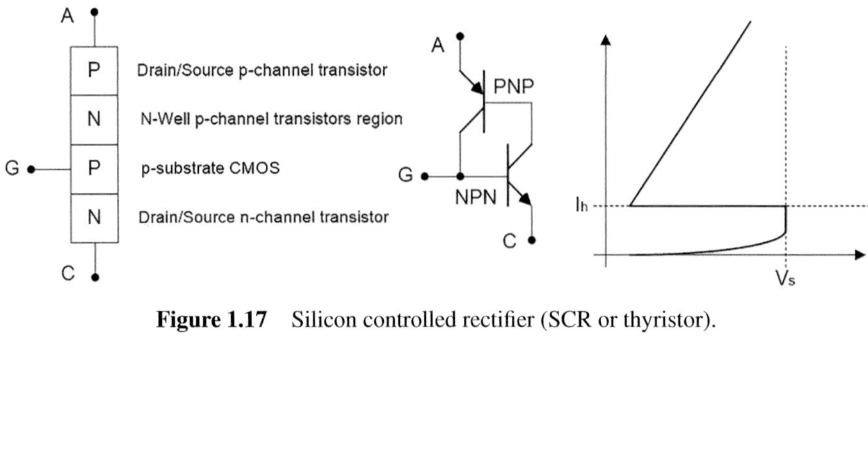

Figure 1.17 Silicon controlled rectifier (SCR or thyristor).

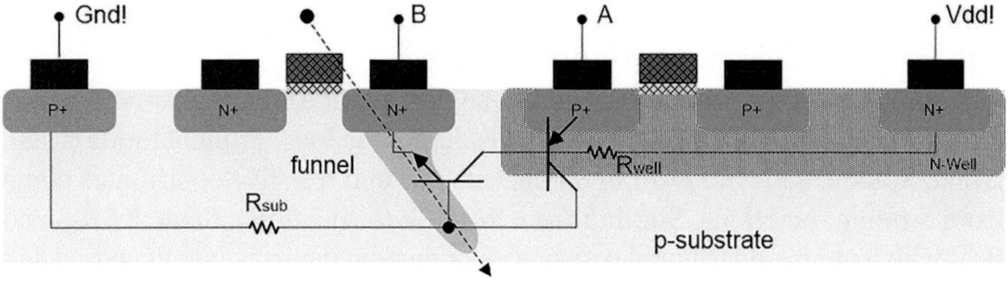

Figure 1.18 Single event latch-up (SEL).

To take back SCR in blocking state, the voltage across terminals A and C (see Figure 1.17) must be reduced to decrease the current flowing in vertical PNP and lateral NPN or, alternatively, the current flowing into the base of lateral NPN must be reduced as well. Once the current is below the holding value, the SCR is off.

In CMOS standard technologies, SCR is usually under control with a good power distribution in the chip and the only way to fall into the "Latch-up" is through power supplies or ground. A spike on these terminals can take the parasitic SCR in the latched state, thus triggering the positive feedback loop and taking vertical and lateral bipolar transistors to sink current. The overall effect is the melting of metal junctions leading to a permanent malfunction.

In space applications HEP generation must be considered, and if charged particles releasing a low energy (low LET) are not responsible of any destructive error, those particles releasing high levels of energy (high LET) can produce a deep funnel region involving both p-substrate and N-Well (Figure 1.18). The generation of an energetic funnel produces a drift current followed by a diffusion current and if the ground strategy of p-substrate is not strong enough, a resistive path (distributed parasitic resistance) can be on the road of the funnel. In this case, the drift current, usually involving the diffusion of the transistor terminal where it comes from, can provide part of its current in the base direction of the lateral NPN transistor, thus increasing the current flowing in the vertical PNP transistor. In this case, a single current spike overriding the hold current of the parasitic SCR takes the region in the latched state and current from power supply takes advantage of the positive feedback loop of the SCR causing the latch-up.

1.4 Single Event Upset: SEU

HEP generation at lower energy (low LET) generates funnel regions not affecting substrate and N-Well (the funnel remains in the region of N+ or P+ diffusion) but still can create drift currents. Such current spikes can create transients (single event transient, SET) consisting in the charge of parasitic capacitances around the N+/P+ diffusion area.

In Figure 1.19 is shown the small-signal circuit for an n-channel transistor in a standard CMOS process [4]. Considering the four-terminal mode (Gate, Drain, Source, and Bulk), different capacitance and resistive contributes come from different junctions. Starting from the bulk (terminal B), there are R_{BS} and R_{BD}, which are the distributed resistance connecting the substrate to source and drain regions; C_{TS} and C_{TD} are depletion-region capacitances at the reverse-biased p-n junction between substrate and source and substrate and drain, respectively. R_S and R_D are the resistances coming from the contact terminal (drain and source) and the channel of the transistor. On the gate side, there are four parallel capacitances: C_{GD} and C_{GS} are inherent to the transistor as a device, while C_{RD} and C_{RS} are parasitic elements coming from the manufacturing flow and are becoming lower and lower thanks to scaling of the technology node and more precise masking tools. C_{GD} and C_{GS} are the capacitance between gate and drain/source terminals and are strongly related to the bias condition of the transistor and its width and length. C_{RD} and C_{RS} represent the capacitance coming from the overlap between the gate and drain/source diffusions.

According to the small signal equivalent circuit, it is now possible to evaluate the effect of a charge particle striking even at low energies (low LET). The funnel region creates a drift and immediately after a diffusion current basically flowing in the N+ region (see Figure 1.20).

The current generated sees several capacitances on its path: C_{TD}, C_{RD}, and C_{GD} (according to the example shown in Figure 1.20). Being a low energy strike, the contribute of C_{TD} is negligible, but C_{RD} and, in particular, C_{GD} must

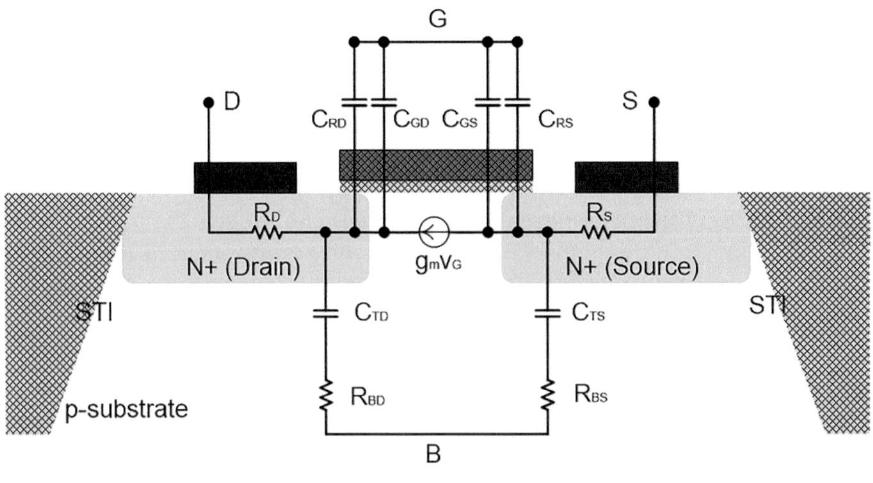

Figure 1.19 Small-signal equivalent circuit for a CMOS n-channel transistor.

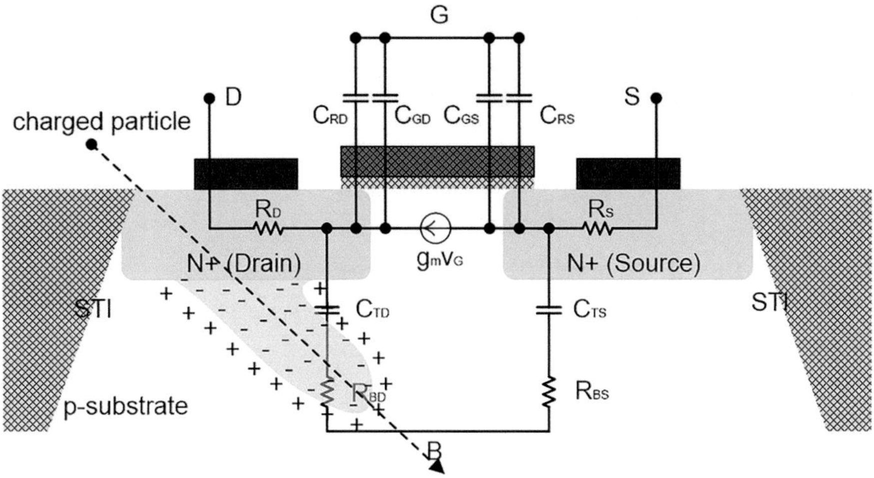

Figure 1.20 Charge particle strike in a CMOS n-channel transistor.

be considered. The charge (or discharge of these capacitance) is responsible for the transient on node D; if this node is well-biased, the effect is negligible, but in case of a specific connection (e.g., a bistable circuit or a diode connection in a current mirror), the result can be an upset or an amplified current spike triggering a cascade effect in a digital flip-flop or in an analog component.

The incident angle of a charged particle is a stochastic event and the hitting may involve more diffusion area such as that shown in Figure 1.21. In this case, the drift and diffusion current may come from two parts and involve

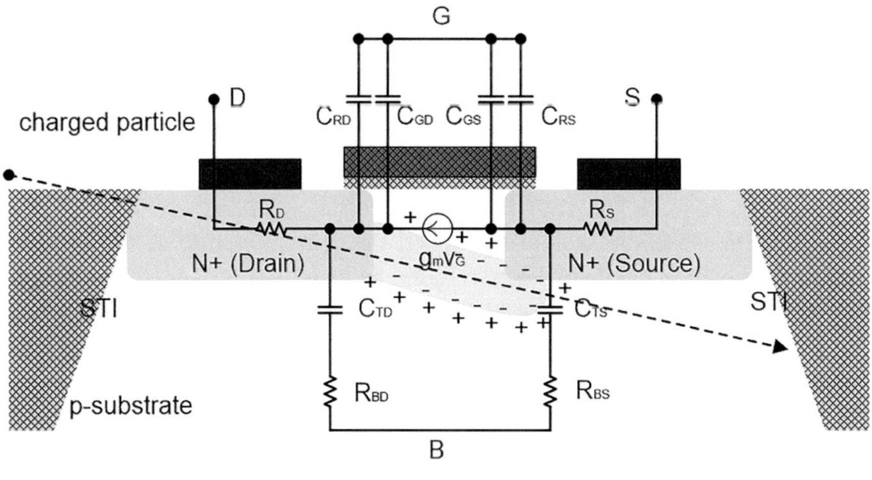

Figure 1.21 Charge particle strike generating a funnel region including drain and source areas.

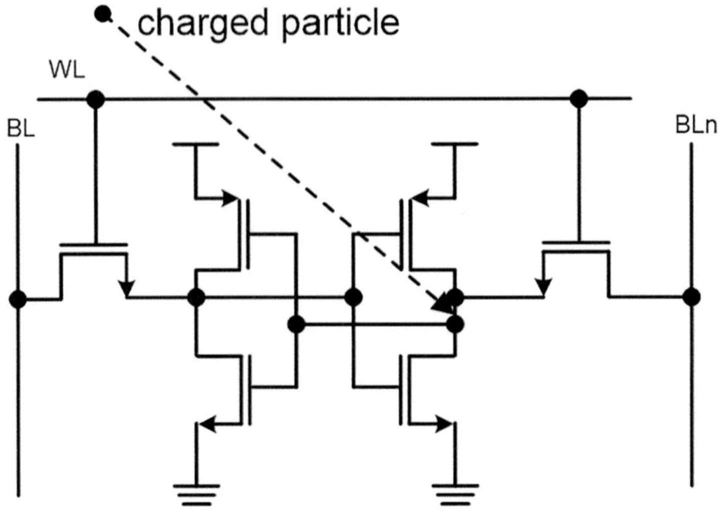

Figure 1.22 Charged particle strike on a bistable node in a 6T SRAM cell.

more capacitances, giving more complex charge and discharge phenomena able to trigger undesired transients.

In Figure 1.22 is shown the hitting of a particle strike on one of the two bistable nodes of a 6T SRAM cell. These nodes are opposite to each other and can be only at ground or power supply. The great advantage of this circuital topology is that in both states the structure has no current flowing along the p-channel and n-channel transistors. The only currents flowing are those able to compensate the leakage currents from transistor junctions, which are negligible in the overall economy of the device. The stability of the bistable pair is guaranteed by capacitors (see Figure 1.21), and to change the state, a voltage can be applied to one (or both) node to make the bistable flipping in the opposite direction. This voltage charges capacitors, and this charge is also called critical charge. Once overrode the critical charge the bistable circuit, thanks to its positive feedback loop, changes state. When the energy release of the charged particle is able to exceed the critical charge of the bistable circuit the positive feedback loop triggers the change of its state (from "0" to "1" or vice-versa).

A charge particle releasing energy generates drift and diffusion currents; if these currents can charge capacitances, keeping the bistable in the previously programmed state overriding the critical charge, we have an upset (a logic "0" becomes a "1" or vice versa); if currents are not enough, we simply have a transient.

1.5 From SET to Soft Errors

As mentioned in the beginning of this chapter, the energy released by an energetic particle becomes a charge injected in transistor nodes having intrinsic capacitance related to the device itself or misalignment of the masks in the manufacturing flow. This event is also called fault injection and the basic assumption is that every node in the net has a specific critical charge that we can also see as the capability to absorb the energy released by the particle. For very low energetic particles, the effect is negligible, but when LET increases, such nodes start showing glitches.

In standard logic ports, such as inverters, NANDs, or NORs, a particle strike can produce a temporary (and very short in time) voltage spike; this spike can be considered as a disturb, and if the architecture of the logic device is robust enough, the disturb may be filtered by using the same methods commonly adopted in digital electronic devices (filters, good disoverlap strategies, etc.).When energy becomes higher, the spike can become relevant in terms of dynamics, and the output signal (see Figure 1.23) can change state for a while before recovering to the original value.

According to the amplitude of the transient, the SET can remain in the port hit by the particle or it can propagate to next cell (see Figure 1.24). In

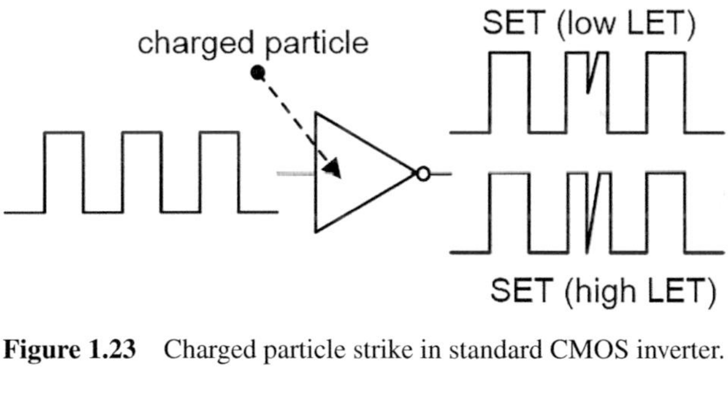

Figure 1.23 Charged particle strike in standard CMOS inverter.

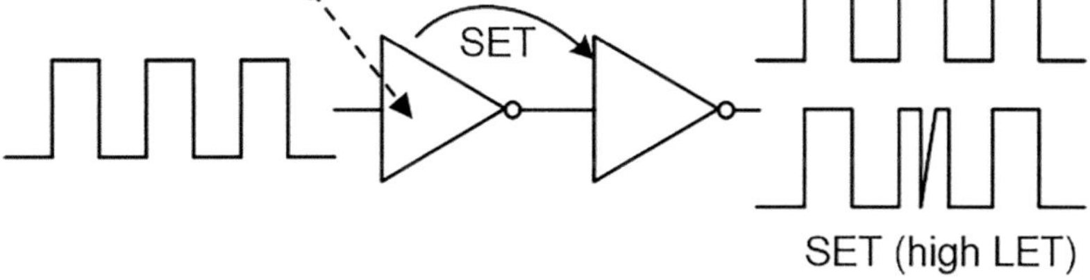

Figure 1.24 SET propagation in a logic chain.

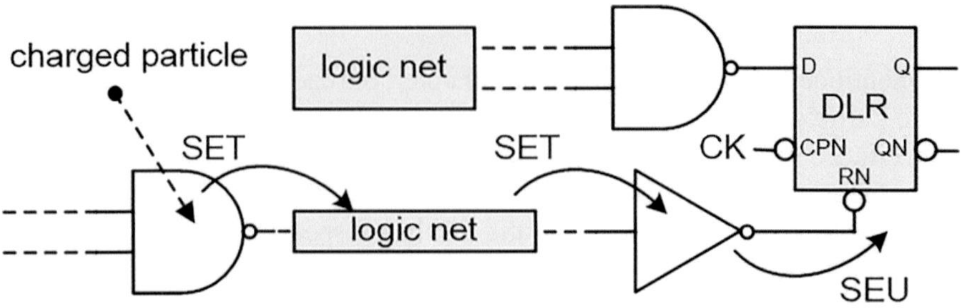

Figure 1.25 SET propagation into a delay latch with reset (DLR) becoming an SEU.

this case, it is possible to have a cascade of glitches running through logic into critical parts of the chip.

In Figure 1.25 is shown how an SET originated by a charged particle can run into a logic net (a set of simple logic ports), and once it arrives at the reset terminal of a latch, it can change a logic value expected to remain high according to the logic function of the digital block. The SET has become an SEU (single event upset) and a simple glitch has been promoted to a soft error.

The example of Figure 1.25 can be moved to more complex architectures, and the soft error can make a second cascade involving more storage elements shaped as a state machine. In digital design (but also in analog design if we consider an ADC or DAC driven by counters), it is quite common to have state machines programmed to have a specific number of states where functions are performed; sometimes not all possible states are considered (to reduce the number of digital cells and the overall complexity of the digital module), and in this case, they are known as finite state machines (FSM). An SEU can take the FSM in a forbidden state where only a reboot or a restart of the digital engine can take out from this stuck condition. This kind of error is also called single event functional interrupt (SEFI).

In Figure 1.26 is shown an SEFI occurring to a counter reading a memory (SRAM or PROM) due to an SEU on the most significant bit (MSB); in this case, if the particle strike comes when MSB is “0” taking it to “1,” a relevant number of addresses are skipped and loaded code is wrong. On the other hand, if the hitting takes from “1” to “0,” the counter comes back to a previous address and read again the same set of addresses giving in any case a failure in the code loading procedure. In this case, even if the digital engine does not enter in a forbidden state and stuck condition is not reached, a functional interrupt (SEFI) comes from the wrong code loading.

In semiconductor memories, the arrays containing memory cells (6T SRAM or non-volatile) are usually organized to minimize area consumption;

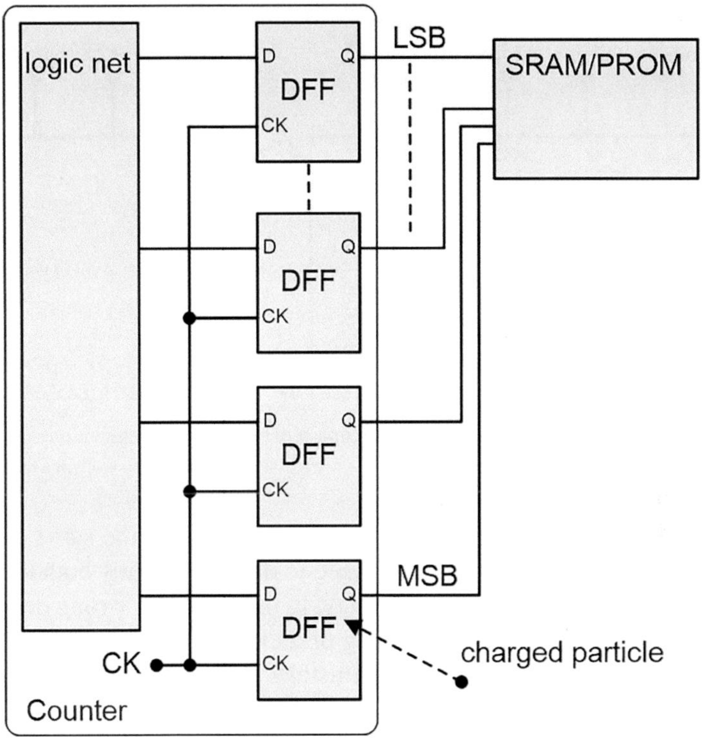

Figure 1.26 SEU in a counter scanning a memory (SRAM/PROM) for code loading.

this means that along a word-line, many cells (bits) share the same wire. In Figure 1.27 is sketched a typical memory organization. When the memory

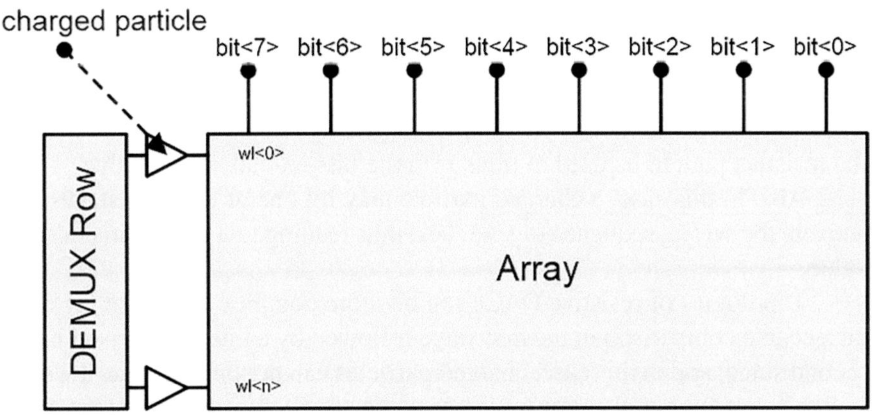

Figure 1.27 MBU from a charged particle hitting a row decoder final stage.

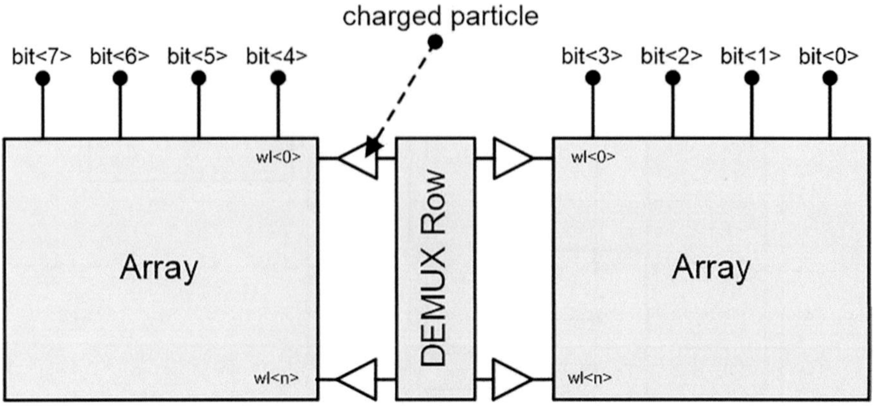

Figure 1.28 MBU on four bits in a separated array organization.

cut is small enough, all the bits building the byte share the same word-line and final stages (buffers) are usually sized to drive efficiently both the closest and the farthest bit minimizing any delay. In this case, the wrong decoding in the DEMUX stage or the direct hitting of a charged particle on the terminal stage driving a word-line can make a mistake along all the bits: this is called a multiple bit upset (MBU).

In large memories, the array can be split and the common DEMUX engine drives two word-lines (left and right) with separated drivers (Figure 1.28). In this case, if the charged particle hits a driver, the result is that four bits are wrong, but still if the particle hitting occurs on the DEMUX engine, a wrong word-line can be decoded, thus leading to an MBU on all the bits of the byte.

Moving from a strictly digital example to an analog one, in Figure 1.29 is sketched a very simple resistive DAC using a demux to drive a 3-bit Kelvin divider [5]. Just like a memory, this DAC includes a part decoding a digital word (3 bits in this case), providing the right signals using buffers to select the resistive path to be used to transform the bits into an analog value.

Also in this case, a charged particle may hit one of the final stages and activate the wrong sequence of switches, thus leading to a wrong final analog value.

Topologies of resistive DACs can be more complex and make leverage on a coarse comparison in the first stage followed by a fine comparison in the second stage, and in this case, charged particles can not only provide a wrong switch selection but also impact the sequence itself during the passage from the first phase to the second.

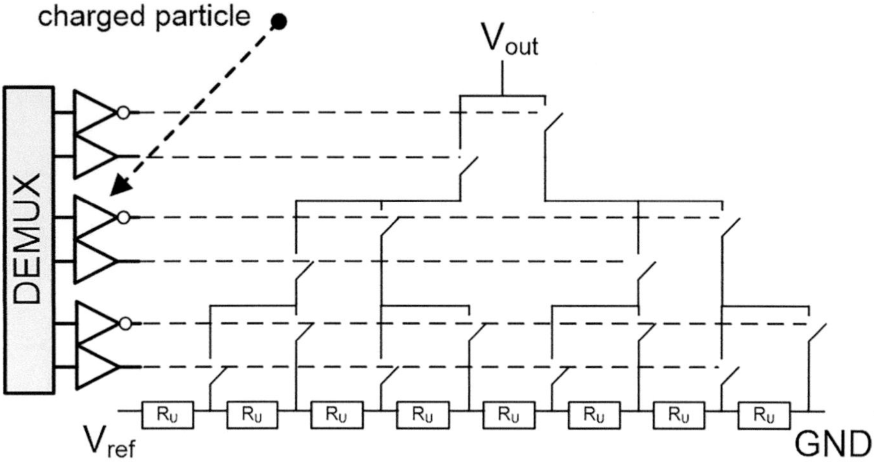

Figure 1.29 Charged particle effect on Kelvin divider in a resistive DAC.

Almost all data converters are driven by a clock signal representing one of the most critical parts both in digital and analog applications. As mentioned above (see Figure 1.24), a clock may be hit by a charged particle with an SET able to reach a sequential port that can change its internal value.

Clock is usually distributed inside the chip using the so-called clock tree. In digital design, CTS (clock tree synthesis) is one of the most critical parts of the overall design, and well-balanced clock paths have to provide synchronized signals in all parts of the chip [6].

Data converters usually do not have complex clock trees, but paths' balancing is even more critical than in digital applications to avoid any mismatch in the core parts of the analog-to-digital conversion. In SAR ADCs, in particular, every step is clocked and a cascade of errors in the clock tree (transients' cascade) can trigger a cumulation of non-linearities affecting strongly DNL and INL and the overall performance of the ADC.

In Figure 1.30 is sketched a CTS example where a charged particle can hit randomly in one of the buffers propagating the clock into the chip. Since CTS in data converters are usually designed by hand (using Place and Route tools for small CTS can be time wasting), a good strategy should be including not only balanced buffers with proper driving strength but also filters and other passive elements (e.g., delay elements) to make the critical charge on critical nodes higher in order to reduce the probability of SET coming from lower energy charged particles.

Delay elements are commonly used also in automatic CTS engines to adjust setup and hold times, but in this specific case, they can also be used

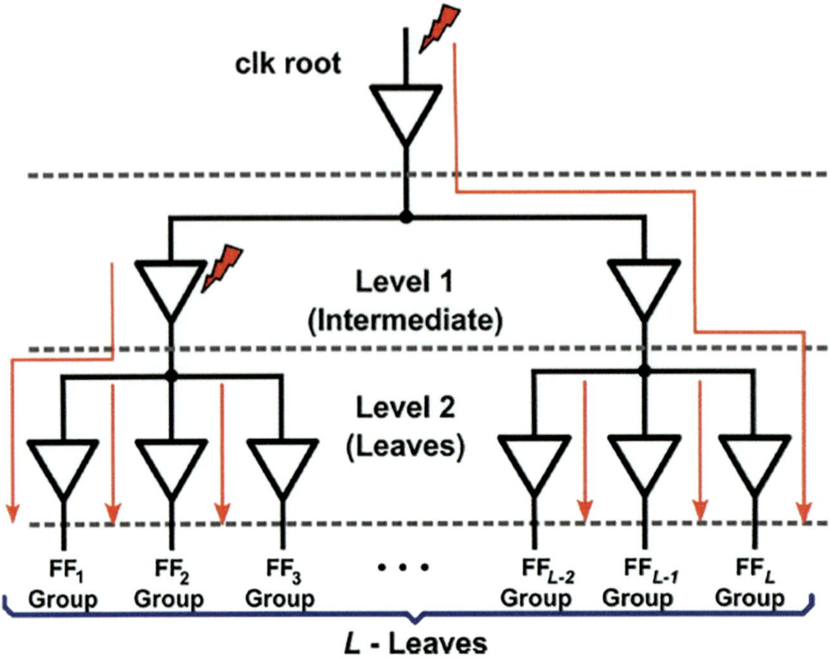

Figure 1.30 Effects of charged particles on a clock tree [6].

(even if unnecessary in terms of balanced propagation) to mitigate the probability to have SET propagation stopping transients in critical parts of the tree.

1.6 Radiation Hardening by Design (RHBD)

Radiation hardening by design is a set of techniques focused on the mitigation of the effect of radiation on semiconductor components. These effects cannot be completely avoided and even by using heavy shields, some effects still occur. Just like occupational health and safety for men at work, also for semiconductor components in hostile environment, the risk of a failure can be only reduced at its minimum but not completely avoided. So, like men at work in a building site, even if protected by the right equipment and following proven procedures, semiconductor devices can fall into an accident.

The RHBD approach is very similar to occupational health and safety procedures. Here the main task is to reduce the probability of an accident at its minimum, keeping in mind that it can always occur even in the best conditions; the second effort is to reduce at its minimum the effects of an accident happening. RHBD techniques firstly consider the destructive events (SEL,

TID) and reduce the probability they can occur with architectural strategies and by using a dedicated physical design (layout); once reducing the probability of such events, the second task is to mitigate the effects at its minimum.

There is a third task that is very well known in the space community, and it is represented by EDAC (error detection and correction). It is a sort of cure of radiation effects, but it is not (and must be not) considered as a panacea. As always happens, EDAC gives the best when a solid architecture behind a good layout design give few errors to be corrected; expecting EDAC resolving all issues coming from components not conceived to be rad-hard is risky.

RHBD has been considered in the past as a handcrafted activity inside the semiconductor market. Nowadays, this is only partially true.

The progressive scaling down of modern technology is taking transistors in the few nanometers area with very reduced power supplies and, of course, very low critical charges in bistable elements. Also, hard errors may occur more frequently, and relevant efforts are spent in power strategies and solid ground connections. In the nanometer era, semiconductor devices may be prone to particles with energies much lower compared to those space community have to consider (heavy ions, protons, gamma ray), and as a matter of fact, aircraft producers are starting to make strategies for electronic equipment able to resist to some kinds of radiations. This cannot still be considered as a rad-hard or rad-tolerant approach but maybe an effort toward very-high-reliability: the threat behind this is very similar.

Radiation hardening can be done not only by using a dedicated design but also by making leverage also on the manufacturing process or by using dedicated shielded packages. The first is called RHBP (radiation hardening by process), and the latter RHBS (radiation hardening by shielding).

RHBP represents a very appealing option, and indeed it is used for some very specific applications. Making a dedicated CMOS or BiCMOS, with additional masks or specific manufacturing steps, can be a way to solve the root cause of HEP generation effects both from a TID and SEE point of view, but it has several drawbacks. The first is of course the cost: making a dedicated process for a very low number of components (as usually requested by space market) can be a waste of resources that can be justified only by classified applications (military) or very specific and critical tasks.

But behind the cost, there is also another problem in the RHBP approach. Creating a new "space silicon process" can be feasible, but "maintaining a space silicon process" is very hard and can be the real killer task. As known, silicon manufacturing takes advantage on the number of wafers produced every month; new technology nodes, in the beginning, present a high variability in the process, and it is only with a fine-tuning of the manufacturing steps that yield rises up to 99% as expected in a fine-tuned process. The higher is the number of the wafers produced, the higher is the control the silicon foundry has on the process: this is called maturity of the process.

Now, a dedicated silicon process, like those done in the academy or research centers for exploring some new limits of the manufacture, cannot reach this kind of maturity because the number of wafers will always be low. Space components are sold not in millions of parts but in hundreds or thousands.

Space applications, nevertheless, need a mature process. As known, space community is quite conservative and usually not keen to adopt new technologies. This is true also for silicon process. The maturity of a process is something very welcome from rad-hard component designers or integrators. In some cases, producers prefer using already existing components proven to have a very high reliability and shield them as much as they can. This is RHBS, and the effort is focused on the package that must be light (to meet weight requirements) and able to reduce the LET of charged particles.

RHBS itself does not provide a rad-hard component but what is usually called a rad-tolerant part, and even if they do not reach the same resistance as rad-hard components, they play a significant role in the space market. Mission running at Loe Earth Orbit (LEO) or having a very limited time-frame can take benefit of rad-tolerant components that usually have a lower cost and, on the other side, a higher number of functions (storage capacity if they are memories, number of bits if they are data converters). As usual, it depends on the mission and making a trade-off is the task that engineers have to take into account.

RHBD in the last 30 years has become one of the main assets for space components. The main pillar behind this approach is the use of standard silicon technologies.

As mentioned above, maturity is critical for space applications, and this is the reason why in RHBD, selected process are usually not the most updated but part of the previous waves. Even if very aggressive nodes nowadays guarantee a strong minimization of area (and power consumption), nevertheless, older technology nodes are preferred. Today an "old" CMOS 180 or 130 nm can benefit from more than 20 years of continuous refinements and the high number of wafers manufactured every month leads to yields very close to 100%. This is definitively a positive aspect for applications asking, first of all, high reliability.

Behind the maturity of a process, there is another positive aspect. These old nodes are available from many silicon foundries all around the world, thus opening the possibility to move from one foundry to another.

Space components market represents a thin slice of the overall semiconductor market, but the main characteristic is to be worldwide; that means few components are sold every year and spread almost everywhere.

As a matter of fact, each country has its favorite silicon sources also in consideration of the fact that large programs are triggered by public entities, which are called to use internal sources; if a country has its own silicon foundry, there is a high probability that it will be the provider in the manufacturing flow.

The portability so becomes critical. A space component designed in a CMOS 180 nm and provided to a country must be moved to a different CMOS 180 nm in another country, and if this can be considered an obstacle on the other hand, it takes the overall approach in "safe" condition.

The passage of a space component from one silicon source to another reduces the risk in case a silicon foundry dismisses a process or decides to stop the manufacturing for that market.

Also, national rules (like ITAR, International Traffic Arms Regulation) can be considered a threat for space components, and if a specific device today is manufactured with a standard CMOS process from a foundry that tomorrow is acquired by a larger corporate deciding that this process becomes classified, the risk is to be not being able to have a manufacturing source anymore.

RHBD represents an umbrella for space components. If used well, it can enable business opportunities, but it can also protect by possible changes from silicon sources also considering that space community use these components, once qualified, for many years: what it can be profitable for a foundry today (because in parallel of space components, there are other consumer applications) can become an obstacle tomorrow.

Under these premises of RHBD and in consideration of the main pillars described above, (maturity and portability), it is now possible to enumerate some possible steps for making space components rad-hard starting from a standard silicon process.

Starting from the silicon foundry process design kit (PDK), the first thing to be done is to discard everything that has been designed and that is part of the PDK: from standard cells to I/Os. This can be considered a radical approach, but the main assumption is that everything has been designed by process engineers not for a rad-hard application but for different purposes, and only one failure (e.g., a latch-up in an output pad) is enough to frustrate an entire chip.

A PDK containing only design rule check (DRC) and layout versus schematic (LVS) engines is also called naked-PDK (nPDK), and this is the starting point for making a real rad-hard component.

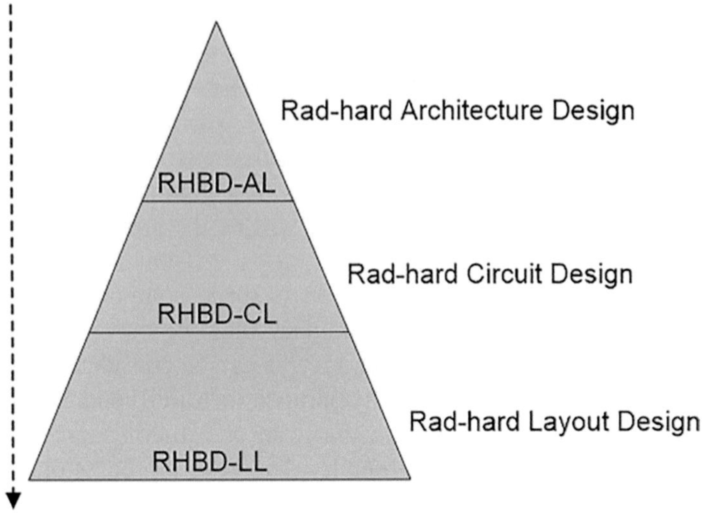

Figure 1.31 RHBD top-down approach.

A good and robust architecture is the starting point for almost every rad-hard component; the right macro-blocks positioning, redundancy, and a preliminary floor plan to guarantee an efficient power distribution and ground lines is the main brick for making easier any successive step in the flow both from the topological point of view (the circuits) and the physical point of view (layout). This is called RHBD at architectural level (RHBD-AL).

In the top-down approach as sketched in Figure 1.31, the next step is represented by the circuit design where topological solutions focused on minimizing the impact of charged particles are considered to avoid hard and soft errors. This is called RHBD at circuit level (RHBD-CL).

The final step is represented by the physical layout design. A strong architectural approach combined with robust circuit solution opens the road to rad-hard devices, but the layout plays a crucial role since the main hard errors coming from TID and SEL are solved with a good and reliable design. This is called RHBD at layout level (RHBD-LL).

1.7 RHBD at Architectural Level

A generic data converter (DAC or ADC) is built by a specific block shaped to be digital or analog. In general, the main block is the array containing the conversion core (a poly matrix resistive or capacitive array in the case of a DAC), while all the circuitry driving the array is called periphery.

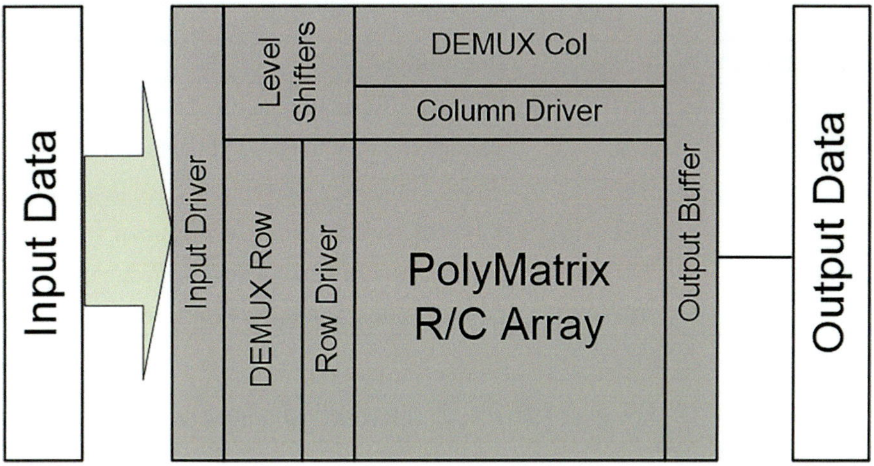

Figure 1.32 Overall architecture of a generic data converter (DAC).

In Figure 1.32 is shown a generic DAC; DEMUX for rows and columns are strictly digital parts; level shifters used to move digital signals from a lower to higher voltage usually has a design closer to canonical analog devices while output buffer provides the analog value out of the chip. The same approach in an opposite way (from analog value to digital output data) can be considered in case of ADCs. Both in DACs and ADCs, different internal architectures (driven or not by algorithms) can be considered; in DACs, the internal matrix (resistive or capacitive can be folded using a coarse and fine conversion), while in ADCs the architecture may be Flash with arrays of comparators or SAR or Pipeline. All these approaches may have some more analog or digital parts; Flash ADCs make leverage on comparators customized for analog purposes and matched in arrays, and SigmaDelta ADCs make leverage on digital structures (filters, counters, etc.).

From an RHBD perspective, all these blocks must be evaluated in terms of hard/soft error making a sort of risk matrix.

From Table 1.1, it is possible to split the architectural effort in two ways: (a) the matrix, where multiple errors must be reduced, and (b) the periphery, where the main threat comes from wrong row/column decoding that can lead to MBU.

SEL, as a hard error, affects all blocks since in all parts, n-channel and p-channel transistors can be close enough to trigger the latched state of parasitic SCR. Nevertheless, level shifters can be an additional threat considering that low-to-high or high-to-low voltage shift may trigger parasitic bipolar paths with latch-ups.

Table 1.1 Risk matrix for DAC blocks.

Macro block	Error	Description
Input/output driver	SET	Risk of SET cascade
Level shifter	SEL	Passage from LV to HV is prone to SEL
Data latching	SEU	Wrong comparing elements selected in the array
DEMUX	SEU/MBU	Risk to decode the wrong row/column
Array driver	SEU/MBU	Final stages can select wrong row/column
Array	MBU/MCU	Disturbs among comparative elements

Figure 1.33 shows an RHBD-AL approach to a generic data converter (a DAC in this case).

From the architectural point of view, the first approach is to tackle the key element of the data convert: the matrix array. There are several ways to convert a digital word into an analog output (using current sources, resistors, capacitors), but considering that active elements are the weakest, a good

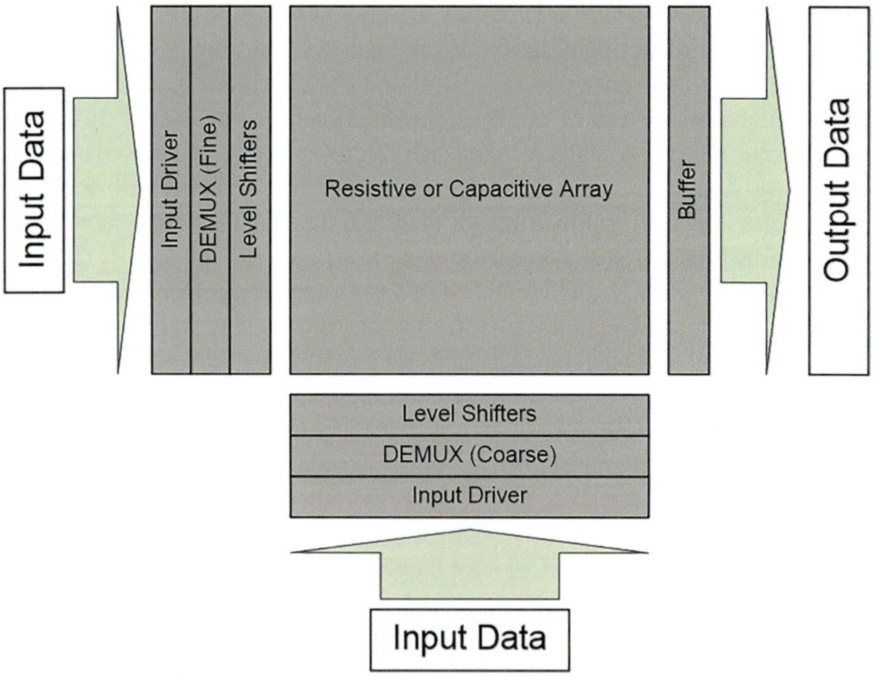

Figure 1.33 RHBD-AL on a generic data converter (DAC).

choice would be using passive elements even if in commercial applications (non rad-hard) they are not the best option because of their weakness in terms of matching during layout design.

Indeed, mirrored current sources can provide a good matching when carefully designed in a layout, but since they are MOS transistors connected as a diode, they can be prone to oxide degradation coming from charge traps at channels' border (intra-device leakage) or between transistors (inter-device leakage). Moreover, mirrored topologies can be prone to SELs because all transistors are usually built using matching elements (stacked gates) not easily protectable with biasing points.

On the other hand, resistors and capacitors are passive and they do not suffer oxide degradation coming from TID. For resistors, it is straightforward mentioning that poly over well is not a good choice (the well can be sensitive to charged particles and oxide boxes may collect charge traps), while poly resistors are immune to TID and SEE showing a good stability even under high charged particles (e.g., Xenon ions).

Also capacitors (in particular, MIM for the same reasons as for resistors) are a good option even if they may have a vulnerability in case of energy release in the array; the effect can be a charge or a discharge during the conversion phase, thus leading to a glitch if the charged (or discharged) capacitive element is involved.

Of course, the adoption of a specific approach to digital/analog or analog/digital brings to a peripheral architecture able to sustain stable boundary conditions like input data without transients and upsets and, as always, hardened from the hard errors point of view.

Regarding decoding schemes, selecting conversion elements inside the matrix, RHBD-AL techniques can be a partial redundancy using voters. In this case, three DEMUXs and a voting logic cell can provide an effective solution: if two decoders among three define a row (because the third has been affected by a charged particle strike), then the result is strictly based on the majority of the votes (Figure 1.34).

For ADCs, considerations are very similar. The nature of the ADC (Flash, SAR, Pipeline, DeltaSigma, etc.) is itself an RHBD-AL point of start; some architecture are intrinsically stronger than others, and, as always, the final application is the main driver. Flash ADCs, strongly focused on comparator arrays, may take benefit of fast conversions having lower probability to intercept transients during the passage from analog to digital; SAR ADCs use more clock cycle for the conversion phase so increasing the probability of a transient passing in critical parts of the conversion. But not only SAR, ADCs can also be considered more "analog" than DeltaSigma ADCs, where

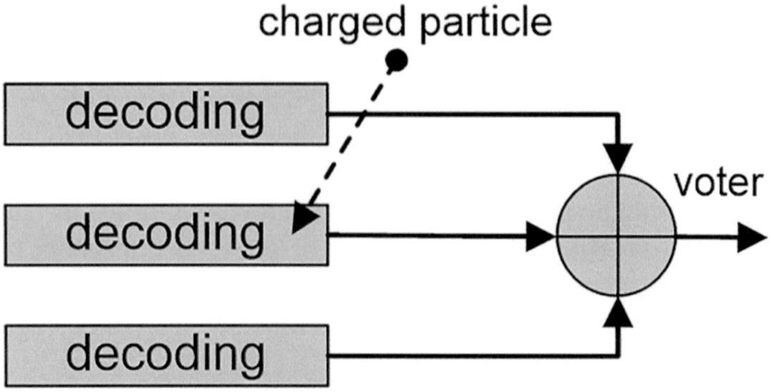

Figure 1.34 Decoding scheme based on voter.

a strong digital approach is used with many standard cells and TID degradation is expected to have an higher impact; on the other hand, the strongly "digital" approach of DeltaSigma make them more prone to transient cascades with high probability that some of them could be transformed into SEU or SEFI, thus bringing in stuck condition the ADC.

Figure 1.35 shows an RHBD-AL approach to a generic ADC using Flash architecture, which, together with SAR, ADC may be considered the most "analog" ADC among all possible architecture (SigmaDelta ADC on the other hand is the most "digital").

In this case, some building blocks can be evaluated from the RHBD perspective and the input stage represented by the sampling phase is one of the most critical. Sampling analog data and holding the sample to be used in the first stage of conversion is prone both to TID and SEE. Even if with a low probability, nevertheless, a charged particle can release energy in the S&H block wasting the sample (a charge collected on a capacitor) and providing corrupted values to the successive elements in the chain (comparators in this case).

As mentioned above for high-energy LETs (e.g., Xenon ions), there is not much to do, but for lower energies (the most common in hostile environments), some solutions can be adopted spreading from the capacitor nature (lower or higher capacitance) and the timeframe for sampling-holding-resampling; the shorter is the timeframe where the sample is held, the lower is the probability that the sample may be corrupted, thus taking advantage to a masking effect from the sampling period. Of course, this approach must consider ADC requirements since playing only on the masking effect can produce other effects in the overall economy of the device.

Regarding TID, the same considerations of the above-mentioned DAC are still valid: the use of passive elements keeping far oxides and its collected

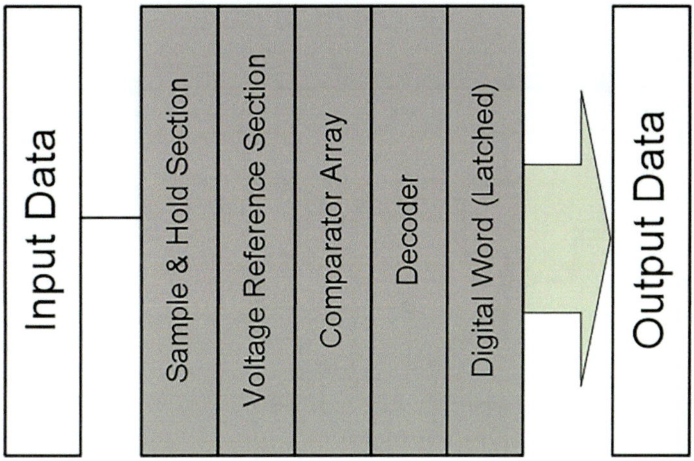

Figure 1.35 RHBD-AL on a generic Flash ADC.

charge traps can be considered and since S&H usually are based on switches, a wise layout hardening of N-channel and P-channel transistors have to be considered to mitigate degradation along MOS borders (intra-device leakage) and between MOS (inter-device leakage).

The Voltage Reference Section is the part where, starting from a positive V_{ref} and a negative V_{ref}, a series of resistance divides the voltage according to the number of bits requested for the ADC; considerations are the same as for DAC: avoiding active elements (poly over active areas) and making leverage on true passive elements like poly resistors in order to be immune against TID.

1.8 RHBD at Circuit Level

RHBD at circuit level is the typical domain of microelectronic designers dealing with single transistors or standard logic cells considering topologies and structures to enhance resistance against radiations. In this domain, the major part of the efforts is spent to mitigate the effects of transients, avoid SET propagations, and reduce the probability to generate SEUs.

In Figure 1.36 is shown a typical RHBD-CL technique for increasing the stability of a 6T SRAM cell when a charged particle strikes one of the two bistable nodes.

As mentioned in the previous paragraphs, the node involved in the hit (Q) has a critical charge mainly coming from gate capacitances and junction capacitances. The positive feedback loop keeps Q and QN in their opposite

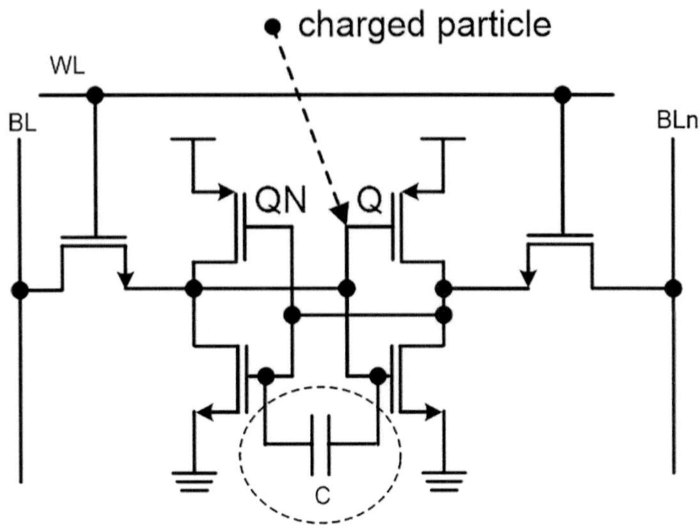

Figure 1.36 RHBD-CL on a 6T SRAM cell.

states, but if the injected charge is higher than the critical charge in that case the feedback loop takes the nodes in opposite direction, thus generating a bit flip.

Since it is a matter of critical charge, one possibility is to increase it by adding a capacitor on both nodes. The capacitor placed on the pairs is called Miller capacitor and can be built using additional poly-silicon structures (poly capacitor) or different metal levels (Fringe capacitor). The result is that a higher energy will be necessary to trigger the SEU.

The same approach can be used also for Latch and Flip-Flop. In Figure 1.37, a D-Latch, having a critical charge represented by the input capacitance of an inverter and the output capacitance of a CMOS switch, can be increased placing a capacitor in the positive feedback loop. This strategy is easier in latches and flip-flops more than in 6T SRAM cells because it is a matter of adding a capacitor possibly shaped as a standard cell. In this case, areas are quite different (6T cells must be at the minimum to build efficient memory arrays) and a D-Latch like the one in Figure 1.37 can be large, but the benefit is that with a good rad-hard standard tool-set and capacitors placed to cut the positive feedback loop, it is possible to obtain structures resilient to SEU up to 60 MeV*cm^2/mg (the energy released by an impinging Xenon ion).

Concerning SETs, they are the typical disturbs occurring to logic ports that, since they do not store any data, can only produce a voltage spike. RHBD-CL, since it is focuse on circuit topologies, can mitigate the amplitude of spice due to energy release in critical nodes and avoid thier propagation

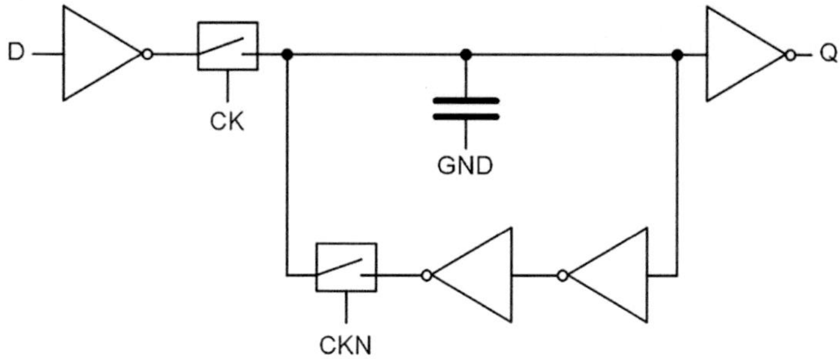

Figure 1.37 RHBD-CL in a rad-hard D-Latch.

along the chain of standard logic cells making a sort of filtering (the transient is stoppped inside).

In Figure 1.38 is shown a typical XOR port using a 6T architecture, which is typical of almost standard cells. In this case, if a charged particle hits one of the transistors not directly connected to the power supply or ground, the reduced dynamics of nodes can amplify the SET, increasing the probability of a propagation in the cell and out of the cell.

The XOR of Figure 1.38 is not rad-hard and placing capacitors on critical nodes to enhance critical charges can represent an overload for a cell not having storage capabilities and with the risk to reduce delays in the propagation of the signal.

A possible solution is shown in Figure 1.39 where in a 10T architecture the number of transistors not directly connected to ground and power supply is reduced at its minimum. In this configuration, the probability to have an SET on such nodes is reduced, and in case it occurs, the propagation is prevented.

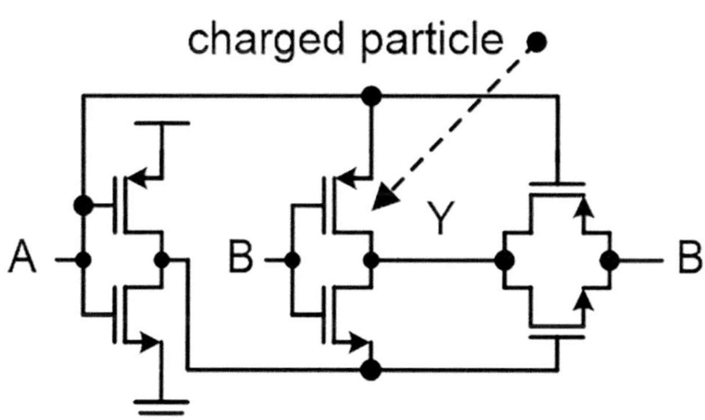

Figure 1.38 XOR logic port using 6T architecture.

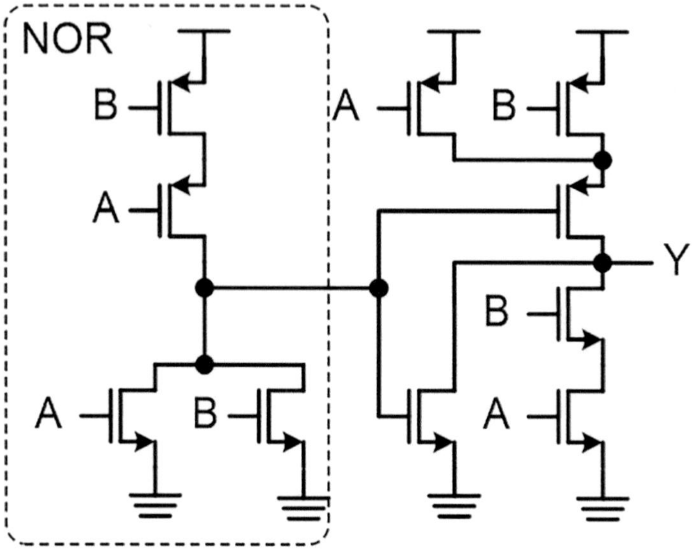

Figure 1.39 RHBD-CL in an XOR logic port using 10T architecture.

From the 10T XOR of Figure 1.39, it is possible to see that this port is the ensemble of a NOR with another elementary cell specific for the XOR transfer function. This approach is quite common in RHBD-CL where the split of functions guarantees an increased resistance to SETs and their propagation.

In Figure 1.40 is shown a typical OR port with three inputs.

The basic block is represented by an NOR cell with three inputs with an inverter in cascade to change the output to the right signal requested by the OR function. This cell suffers from the same problem of 6T XOR port since there are two p-channel transistors not directly connected to power supply and where a charged particle strike can produce SET issues.

Keeping in mind the general rule of maximization of transistor terminal to ground (n-channel) or power supply (p-channel), a possible solution is the scheme shown in Figure 1.41.

In this case, to minimize the number of transistors not direclty connected to power/ground (NOR3 in Figure 1.40) a combination of a NAND, NOR and inverter (Figure 1.41) is used to build the same OR3 cell.

The ensemble of Figure 1.41 takes 10 transistors (four for NAND and NOR and two from the inverter), while the original OR takes eight transistors: the penalty for a rad-hard architecture is two transistors, but the result is a cell resilient to SETs and mitigating propagation of disturbs.

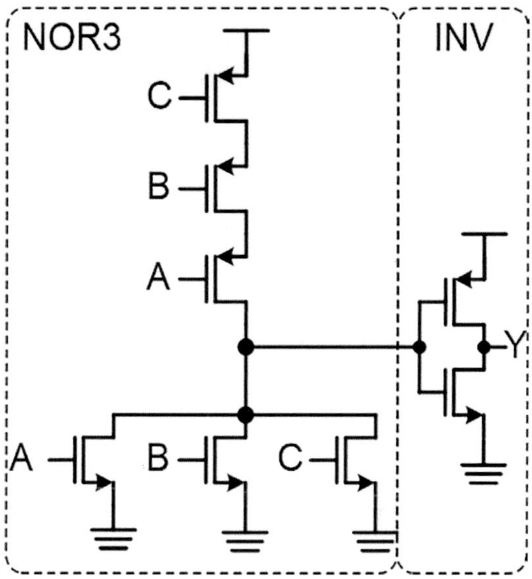

Figure 1.40 OR with inputs as an ensemble of an NOR with three inputs and one inverter.

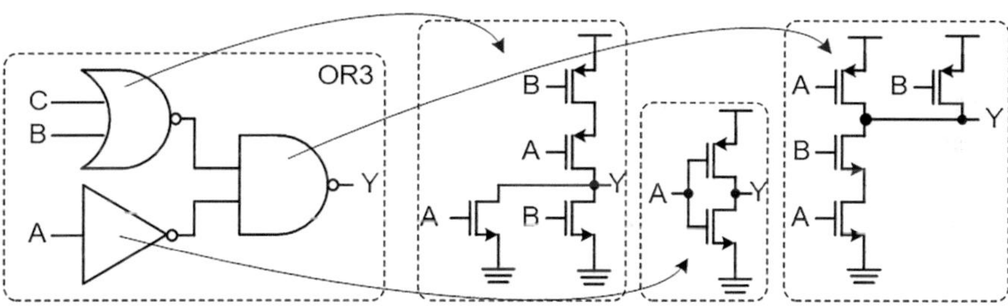

Figure 1.41 RHBD-CL in a three-input OR cell as an ensemble of NOR, NAND, and inverter.

1.9 RHBD at Layout Level

Layout domain is the most critical in the process of making a rad-hard component, and even with a good architecture and a wise topological strategy for single circuitry, the role of a good physical design is of the utmost importance. In particular, hard errors are the ones to be considered at layout level, and we include those coming from TID (degradation of transistor channel) and SEL (parasitic SCR triggered by charged particles).

Concerning SEL, the RHBD-LL shown in Figure 1.42 is one of the most efficient in avoiding the activation of the positive feedback loop

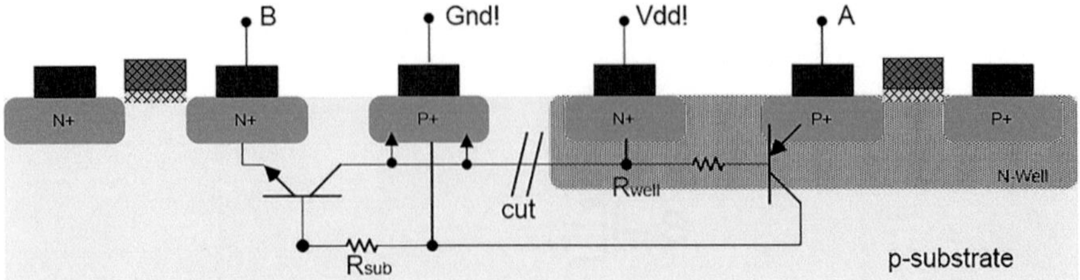

Figure 1.42 RHBD-LL SCR positive feedback loop cut by using enhanced guard ring.

represented by the parasitic SCR in CMOS technology in those regions where n-channel transistors and p-channel transistors are close to each other. In this case, just placing substrate biasing and N-Well biasing in the middle of these regions is enough to create a cut in the feedback loop. The overall result is that the collector of lateral NPN parasitic transistor is connected to the substrate biasing (Gnd!) and the base of the vertical PNP parasitic transistor is shorted to the N-Well biasing (Vdd!). The reverse-biased P+ and N+ regions guarantee that there is not any path to close the loop and even in case of a funnel producing a drift current in the base of the NPN lateral transistor, this is discharged in the substrate biasing terminal connected to ground. The parasitic SCR cannot enter the latched stated as better described by Figure 1.43.

From the design point of view, under these premises, layouts must be designed separating and isolating n-channel regions from p-channel regions (see Figure 1.44).

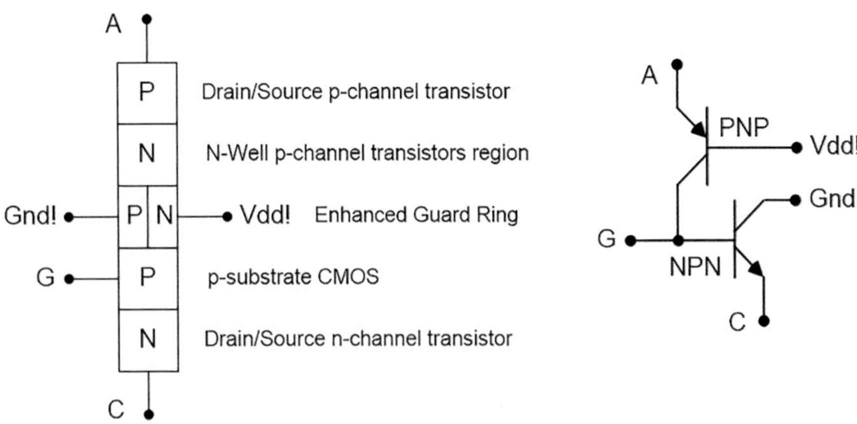

Figure 1.43 Role of the enhanced guard ring in the parasitic SCR.

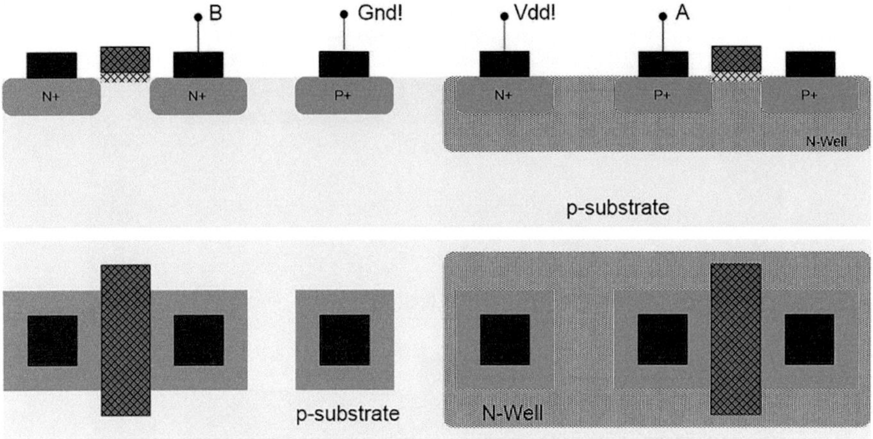

Figure 1.44 RHBD-LL technique to avoid SEL.

This technique has the great advantage to mitigate also TID from an inter-device leakage point of view in consideration of the fact that guard rings prevent any conductive path built by charged trapped in STI.

The second problem that can be solved only at layout level is the degradation of the transistor channel along borders. In Figure 1.45 is shown a common structure used to avoid borders along the channel; this configuration is known as edge-less transistors (ELT) or annular transistors, and it is shaped as a ring having one terminal inside the ring and the second terminal outside. The result is still a transistor and the poly line moving out of the transistor region onto STI still produces a leakage path, but since it is between two regions having the same potential, the result is null. This simple trick makes the transistor never leaky even at very high doses, and it is the only way to maintain the stability of the threshold in every condition.

The intrinsic robustness of ELTs has several drawbacks: on top of all the width that cannot be at the minimum provided by the lithography. Since the ring must be closed around the internal contact, the width will be always very far from its minimum and the same can be said for the length; the angle that in almost all CMOS technologies drawing shapes using 90 degrees angles is not possibile so, to avoid this obstacles, usually 45 degrees can smooth such angles but leading of course to an effective channel lenght that cannot be at the minimum offered by the process.

Another drawback of ELTs is that they are asymmetric devices: drains and source areas and perimeters are different, and this means that in some conditions, the device can work somehow differently from the canonical

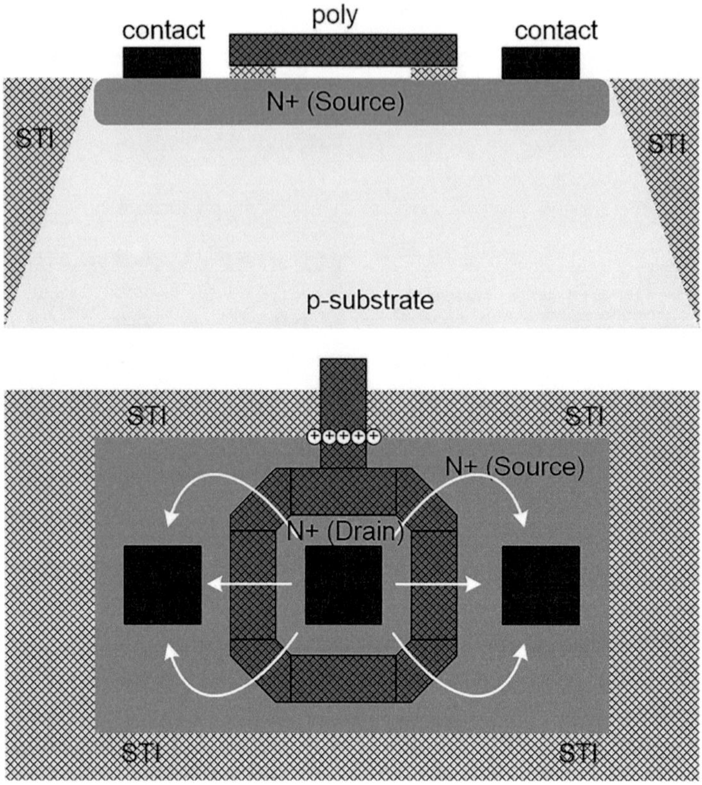

Figure 1.45 RHBD-LL edge-less transistor (ELT) to avoid borders degradation.

transistors. This effect is negligible if the use is digital (inverter, NAND, NOR, etc.) but must be carefully evaluated in an analog design: making a current mirror selecting a drain internally or externally in the ring may give a difference in the scaling (current multiplied by two or more).

The current flowing in the channel is not uniform like in standard transistors. In Figure 1.45 are sketched current contributes when moving from the drain to the source. In this example, also considering the positioning of the source contacts, the channel works in two directions (left and right) with contributes from the angles of the ring. In the top part (where the poly exits from transistor region), the contribute is very low and in the bottom part, if no contact is available in the source region, the same. Understanding which is the real width of an ELT is not easy, and most conventional CAD tools are not able to determine the effective current flowing.

In Figure 1.46 is designed an ELT with an additional contact in the source region. In this case, the channel works efficiently in three directions with minor contributions from the four angles. The fourth side (the one having

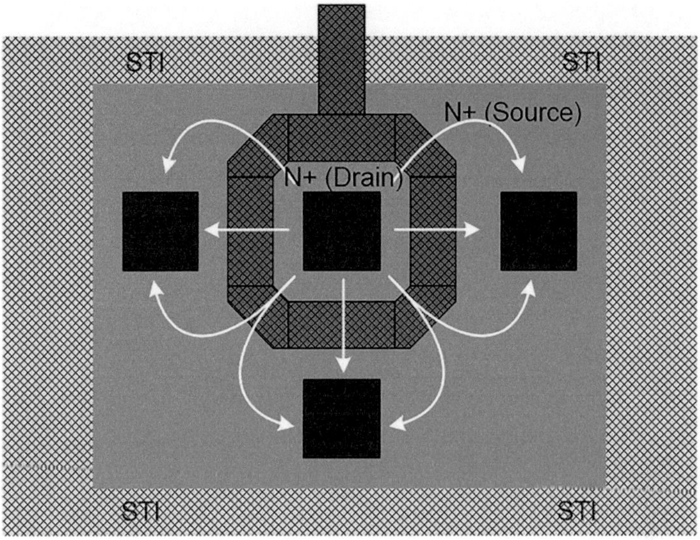

Figure 1.46 ELT with source contacts making the transistor channel more effective.

the connection out of the transistor) still contributes partially to the overall current flow.

When transistor width must be larger than the minimum, a good strategy is always to keep one side at the minimum admitted by the lithography and on that side provide the poly connection of the transistor gate (see Figure 1.47).

In Figure 1.48 is shown a typical example of a layout implementing both P-channel and N-channel transistor. The annular shape avoids the effect

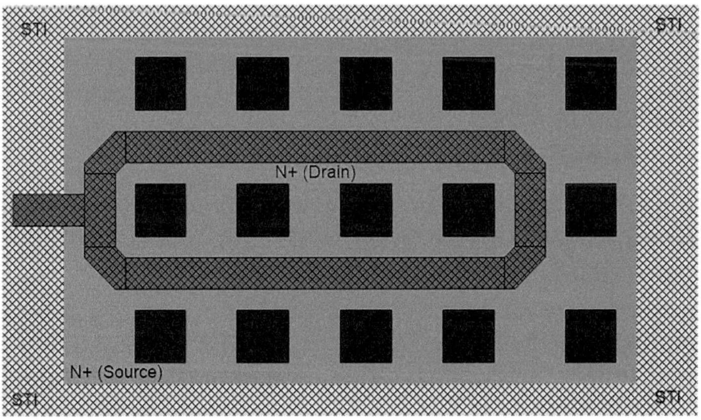

Figure 1.47 Contacts positioning in an ELT with larger width.

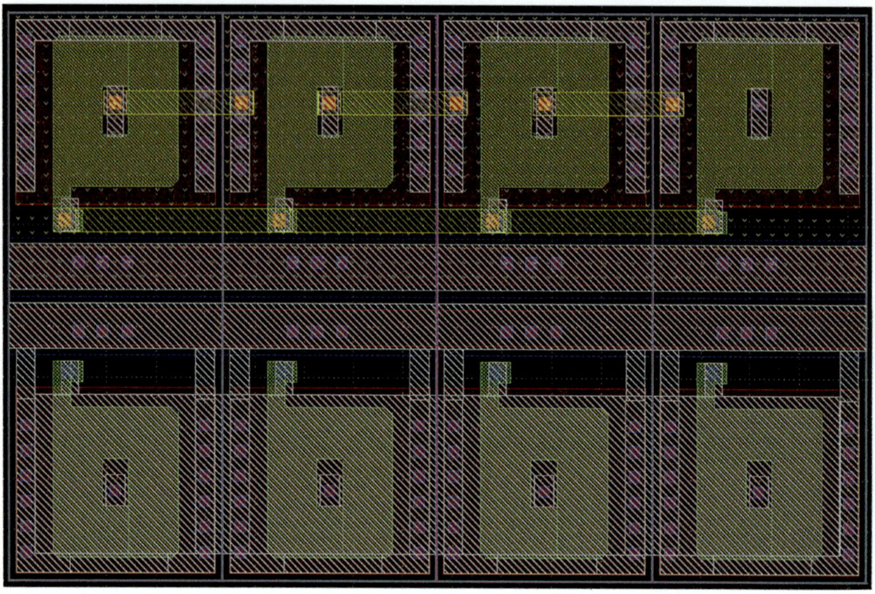

Figure 1.48 Detailed layout of P-channel (top) and N-channel (bottom) ELT transistors.

of the border and the intermediate guard ring separating PMOS and NMOS avoids SEL.

Designing a layout with these constraints can be not so trivial, in particular if an analog approach is adopted. The usual matching obtained using stacked transistors in this case is lost since ELT shape cannot be stacked in any way. Different techniques can be adopted like annular transistors placed in common centroid or long and straight ELTs organized as "organ pipes" to emulate in some way a sort of stacked transistor (Figure 1.49). Of course, this solution is valid only for transistors requiring a large width like those used in power approach to provide large currents.

1.10 Conclusion

RHBD is a set of techniques to mitigate errors coming from charged particle impact on semiconductor devices. Such errors can be divided into hard (destructive) and soft (temporary) and can be the overall result of progressive exposure to radiation (TID) or stochastic (SEE).

Three different domains must be considered when designing a rad-hard component: (a) the architecture; (b) the circuit topology; (c) the physical layout.

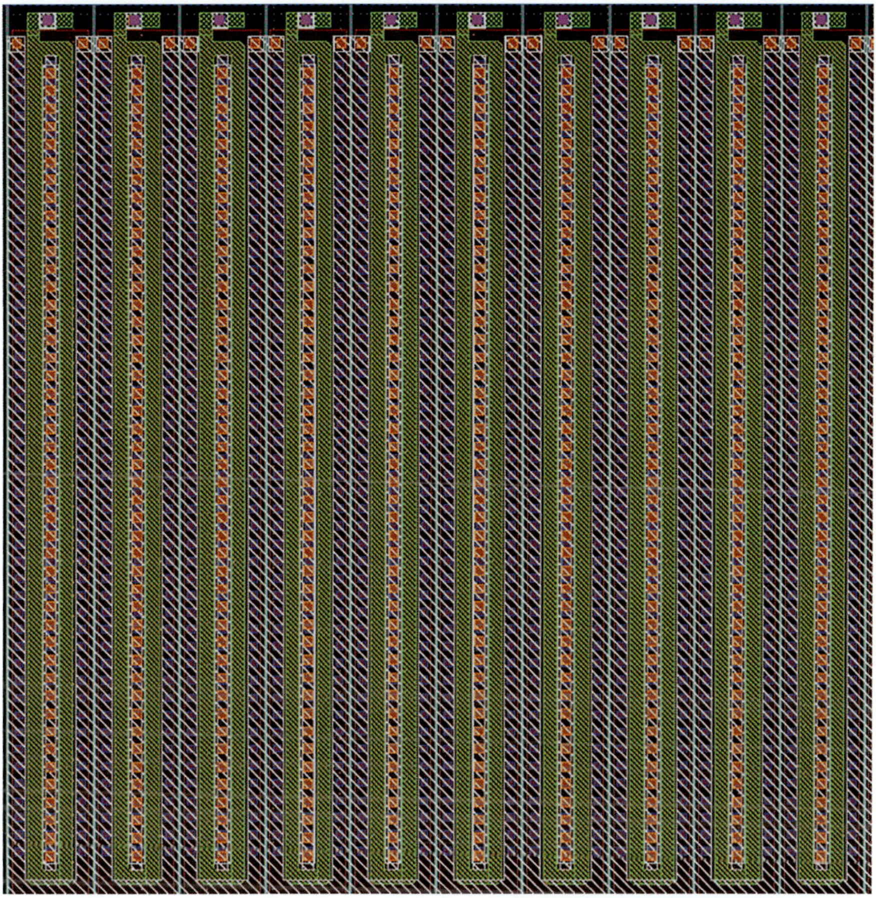

Figure 1.49 ELT transistors organized in organ pipe for power applications.

Each of these domains mitigates specific effects. Architectural domain deals with SEFI and MBU and only with robust main building blocks (e.g., FSM with all possible stated under control or bit distribution in a memory) it is possible to clean functional error bringing, in particular in digital ASICs, in stuck condition. A wise architectural strategy makes easier, in the top-down approach shown in Figure 1.50, the following RHBD steps at circuit and layout levels.

RHBD-CL is in the middle of the architecture and the physical layout and good circuit strategies can play a significant role not only on the circuit but also on the other domain. A robust circuit implementing an already robust function defined at architectural level makes the final device truly rad-hard,

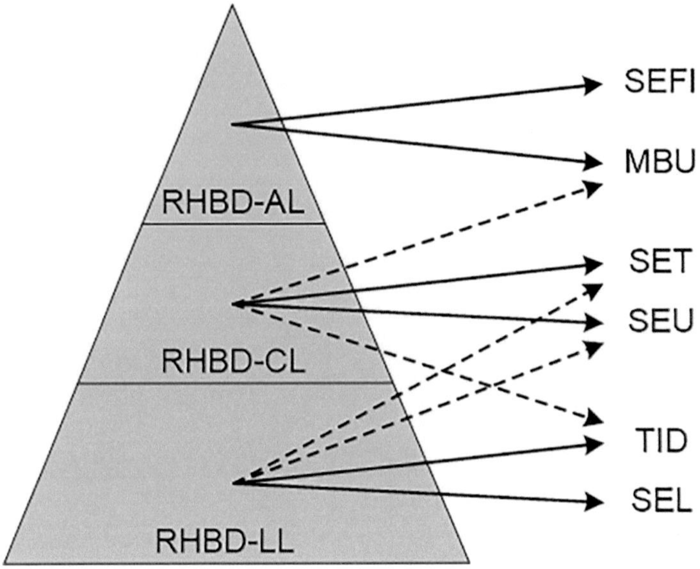

Figure 1.50 RHBD contributes to hard and soft errors in different domains.

not forgetting that a "simple" circuit can make the physical design easier. RHBD-CL deals mainly with SET and SEU, but MBU mitigation and TID are also in the scope of this domain.

RHBD-LL is the executive arm of architectures and circuits. Since everything must be, at the end, transformed in a physical layout, it is obvious that this domain plays a crucial role. The layout is responsible at 50% of a successful rad-hard device, and this is because it is the one dealing mainly with hard errors. SEL, even if wise topologies are used, requires a robust layout and TID can be mitigated only with layout techniques (separation of n-channel and p-channel transistors, ELT).

Of course, layout cannot solve everything. Too much aggressive architectures (high-frequency clocks, FSMs not completely under control) and circuits (too much compact digital cells) cannot demand to layout only all radiation hardening issues; many compromises must be chosen and balanced, as usual, with the requirements coming from the mission.

References

[1] H. J. Barnaby et Al. "Total-Ionizing-Dose Effects in Modern CMOS Technologies". Transactions on Nuclear Science 53 (2006), pp. 3103-3121. 10.1109/TNS.2006.885952.

[2] A. H. Johnston "The Influence of VLSI Technology Evolution on Radiation-Induced Latchup in Space Systems". IEEE TRANSACTIONS ON NUCLEAR SCIENCE, VOL. 43, NO. 2, April 1996, pp 505-521.
[3] L. Deferm, G. Badenes "Silicon, Local Oxidation of (LOCOS)". Encyclopedia of Materials: Science and Technology (Second Edition) 2001, Pages 8563-8572. https://doi.org/10.1016/B0-08-043152-6/01527-8
[4] R. Muller and T. Kamins "Device Electronics for Integrated Circuits". Wiley ISBN 0-471-88758-7, pp. 441-443.
[5] C. Calligaro and U. Gatti, "A CMOS 12-Bit 3MS/s Rad-hard Digital-to-Analog Converter Based on a High-Linearity Resistor String Poly-Matrix", Chips 2024, 3, 129–152. https://doi.org/10.3390/chips3020006
[6] J. J. Saenz Noval, U. Gatti and C. Calligaro, "A Comprehensive Methodology for Soft Error Rate (SER) Reduction in Clock Distribution Network", Chips 2025, 4, 39. https://doi.org/10.3390/chips3020006
[7] C. Calligaro, V. Liberali, A. Stabile "Design of a Rad-hard Library of Digital Cells for Space Applications", 15th IEEE International Conference on Electronics, Circuits, and Systems.

2

Rad-hard Voltage References

Abstract

Accurate voltage references find application in virtually all electronic circuits and systems, including analog-to-digital (ADC) and digital-to-analog (DAC) converters, flash memories, temperature sensors, and amplifiers, thus being key components for many space exploration applications. The performance of a voltage reference has a significant impact on the performance of all the subsequent circuits, which depend on an accurate and stable bias voltage/current. In such cases, having reference circuits that provide very precise outputs under any extreme condition (like under radiation exposure) is mandatory to avoid mission failures. In this chapter, first the basic concepts of general voltage references are summarized (voltage/current mode, first-order compensated, etc.), while in the following, we will review the main solutions (architectural and physical) presented for designing in a standard CMOS, namely class of circuits as rad-hard, with respect to signal quality and microelectronics survivability. The focus will be mainly on TID contrast, while a shorter mention of mitigation techniques against ASET will be illustrated. Then, a look will be given to new materials which revealed as promising. Finally, two rad-hard bandgap voltage references (fixed and trimmed) recently implemented by the Authors are described in detail. Each of them is implemented in three versions with different output voltage (0.6 V, 0.9 V, and 1.2 V).

2.1 Introduction

Voltage references are among the most popular basic building blocks in CMOS complex systems and circuits. They are essential to generate an accurate voltage, V_{ref}, from which all the other references (either voltage or currents) inside the system can be derived. Stable current references are then

DOI: 10.1201/9788743813286-2

often derived by V_{ref}. It is worth mentioning that good transistor biasing is crucial since both static and dynamic performance depend on it. They are used in data converters, regulators, bias generators, and sensor interfaces. Any fluctuation in their output, due to radiation, compromises the performance of the full system. So that V_{ref} and I_{ref} shall be as independent as possible from supply, temperature, and process variations (PVT). An environment full of radiation adds one more cause of performance degradation, which shall be overcome. In this chapter, we focus first on basic concepts and general design solutions, which will be the starting point for the development of the rad-hard counterparts, described in the following sections. The chapter continues with a brief view on an emerging technology (i.e., SOI, SiGe, etc.) applied to such a class of circuits, which, although not yet mature enough, could be a winning answer for future and more demanding developments. Lastly, two rad-hard bandgap voltage references (fixed and trimmed) developed by the Authors are described.

2.2 Voltage References Basic Principles

Precise voltage references are built up starting from accurate elements available in a standard CMOS technology. In the following, common overview material and basic principles behind their operation will be reviewed, focusing on the most popular topologies and on temperature dependence. Other performance, such as line regulation, PSRR (Power Supply Rejection Ratio), etc., will not be treated here. A complete description of all characteristics of these circuits is outside the scope of this book and can be easily found in existing literature [1]–[3].

Generally, voltage references insensitive at first order to temperature are obtained by combining two different sub-circuits that are proportional to absolute temperature (PTAT) and complementary (or conversely) to absolute temperature (CTAT), respectively. Depending on the "core" element used to generate PTAT and CTAT, the family of the voltage reference can be divided into two branches:

- Parasitic diode-based (p-n junction) references
- MOSFET-based (voltage threshold) references

In a CMOS process, the electrical characteristics of the parasitic p-n junctions are of fair quality since they are not strictly controlled during fabrication, so that sometimes MOSFET-based voltage references are preferable.

2.2.1 Parasitic diode-based voltage references

Starting elements which generally constitute diode-based voltage references are:

- Base-emitter voltage V_{BE} of the parasitic bipolar transistor
- Thermal voltage $V_T = kT/q$ (about 25.8 mV at room temperature)

where T is the absolute temperature, k is the Boltzmann constant (1.38×10^{-23} J/K), and q is the electron charge. At a given temperature, their accuracy is in the order of a few percent.

One of the most common approaches to the generation of a voltage reference derives from the base-emitter voltage V_{BE} of a bipolar pnp transistor in a diode configuration. This device is an intrinsic parasitic element of the CMOS technology. Its simplified cross-section is shown in Figure 2.1, where the vertical BJT is made up of the p+, n-well, and p-sub. Because the substrate is connected to ground, the diode's cathode (K) must be tied to ground.

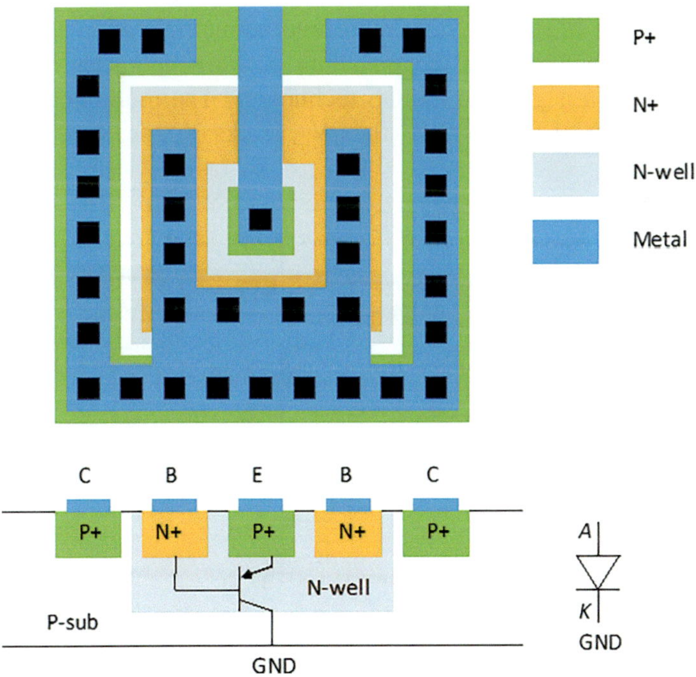

Figure 2.1 Top view and cross-section of a vertical pnp transistor in CMOS technology.

When the diode is forward biased, the temperature dependence of V_{BE} is negative (in the order of roughly –2.2 mV/°C at room temperature, depending on the DC current through the diode), since it is related to the increase of electron-hole generation with T. Whilst the thermal voltage V_T has a positive temperature coefficient (0.086 mV/°C).

The base-emitter voltage V_{BE} is rather accurate and can be used to generate a voltage reference by suitable multiplication. Figure 2.2 shows a typical topology, called diode-referenced self-biasing.

The circuit works as follows: Wilson mirror made up of M1 to M4 keeps the current I in the two branches equal, so that voltage $V(1) \approx V(2)=V_{BE}$. The current I is, in turn, equal to V_{BE}/R. The reference voltage V_{ref} is therefore given by

$$V_{ref} = kRI = k\frac{\left(\frac{W_5}{L_5}\right)}{\left(\frac{W_2}{L_2}\right)}V_{BE} \tag{2.1}$$

where k and M5/M2 aspect ratio represent the multiplication factor for V_{BE} and are dependent only on the matching between geometry quantity and resistance, which is good in CMOS technology. The multiplication ratio is also independent of temperature, thus V_{ref} tracks V_{BE} negative temperature

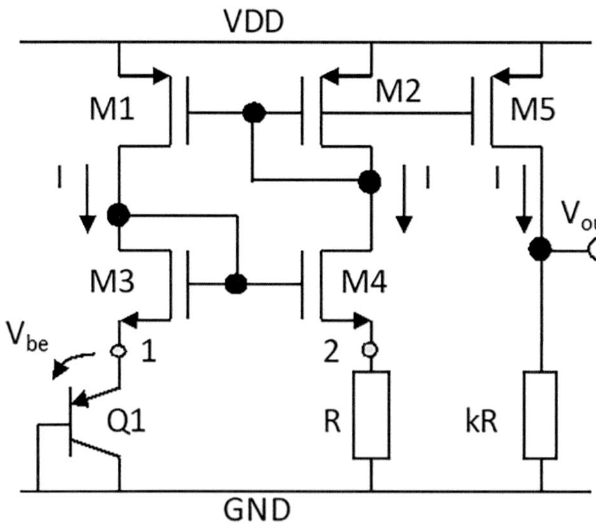

Figure 2.2 Example of voltage reference generated by V_{BE} multiplication (CTAT) [1].

dependence. Since V_{BE} decreases about −2 mV/°C at room temperature, V_{ref} has a negative temperature coefficient.

Another option to design a voltage generator is to leverage the thermal voltage V_T. In this case, we obtain an output voltage directly proportional to absolute temperature. An example of such a circuit is shown in Figure 2.3. Two pnp diode-connected bipolar transistors have a ratio of 1 to n between their areas. The voltage ΔV across resistance R is given by V_{BE1}-V_{BE2}, and assuming that M1 to M4 are matched (same current flows in two branches), it results in

$$V_{BE1} = V_T ln\left(\frac{I}{AI_{SS}}\right) \tag{2.2}$$

$$V_{BE2} = V_T ln\left(\frac{I}{nAI_{SS}}\right) \tag{2.3}$$

where A is the area of the emitter of Q_1 and I_{SS} is the reverse saturation current. Thus, we get

$$V_{ref} = kRI = k\Delta V = kV_T ln(n)$$

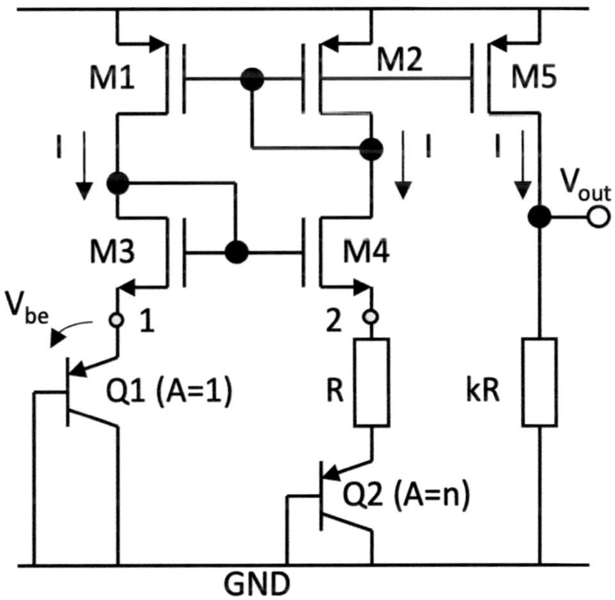

Figure 2.3 Example of voltage reference generated starting from V_T (PTAT) [1].

Thermal voltage V_T has a positive temperature coefficient. Thus, being k and n constant, the output voltage shows a positive linear dependence on the temperature, as well.

It is worth mentioning that 8 is a suitable value for n factor, since in a matrix of 3 × 3 vertical BJT, it is easy to lay out a parallel connection of 8 transistors for Q_2 and 1 transistor for Q_1.

Both topologies used in Figure 2.2 and Figure 2.3 suffer from the channel modulation problem, which results in a small difference between voltages at node A and B. To overcome this issue and make V(A) = V(B), it is possible to use an opamp-based equalization technique [3], as will be shown later.

2.2.2 MOSFET-based (voltage threshold) references

As we mentioned before, parasitic vertical pnp bipolar junction transistor (BJT) are not well controlled and are highly process dependent in a CMOS process. Consistent efforts have been made to create voltage references without the help of bipolar transistors.

An alternative solution foresees MOS as starting elements, and the voltage reference V_{ref} is extrapolated in some way from the threshold voltage (V_{th}) of the MOSFET itself. For example, V_{ref} can be derived using the difference between threshold voltages of two MOS transistors [1] or by adding PTAT and CTAT currents coming from MOS operating in the sub-threshold region [2]. While the former solution is achieved by a suitable additional implant step and is not always available, the latter can be obtained with a standard MOS process, similarly to the differential ΔV described in the previous section. Its principle is shown in Figure 2.4.

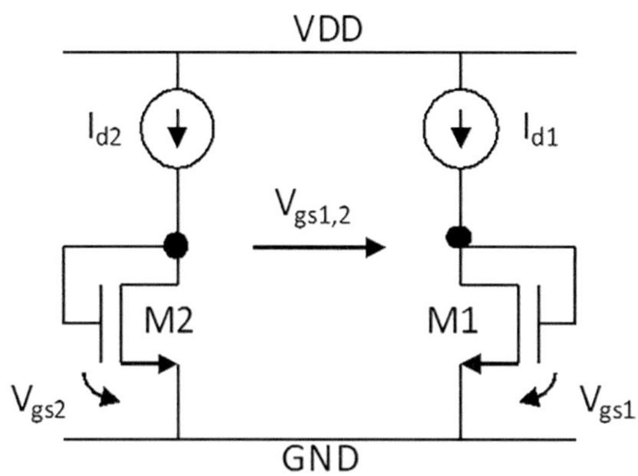

Figure 2.4 Extraction of ΔV_{GS} from two MOS operating in the sub-threshold region [2].

The difference between two MOS having the same size biased in the sub-threshold region with different currents, I_{D1} and I_{D2}, is a function of temperature through V_T and the sub-threshold coefficient ε, as in the following

$$\Delta V_{GS} = V_{GS1} - V_{GS2} = \varepsilon V_T \ln(I_{D1}/I_{D2}) \tag{2.4}$$

where ΔV_{GS} is proportional to T (PTAT), and has good linearity [2].

More complex solutions have been proposed (e.g., in [4]–[5]), as well, all relying on the addition/subtraction of voltage/current coming from MOSFET operating in sub-threshold or triode regions.

2.2.3 First-order compensation

Till now, we showed voltage references that were proportional (directly, V_T, or conversely, V_{BE}) to the temperature. However, in high-accuracy analog circuits, the availability of a voltage reference that does not change with respect to temperature is a crucial point. The approach is to find a way to keep the temperature linear coefficient near zero (first-order proportionality to T), while leaving higher order dependencies as they are. The principle is to compensate the negative temperature coefficient of V_{BE} with the positive temperature coefficient of V_T (Figure 2.5). The compensation can be exploited in two ways: voltage addition (CTAT + PTAT); current addition + I/V conversion.

Voltage V_{ref} can be expressed by

$$V_{ref}(T) = V_{BE}(T) + kV_T(T) \tag{2.5}$$

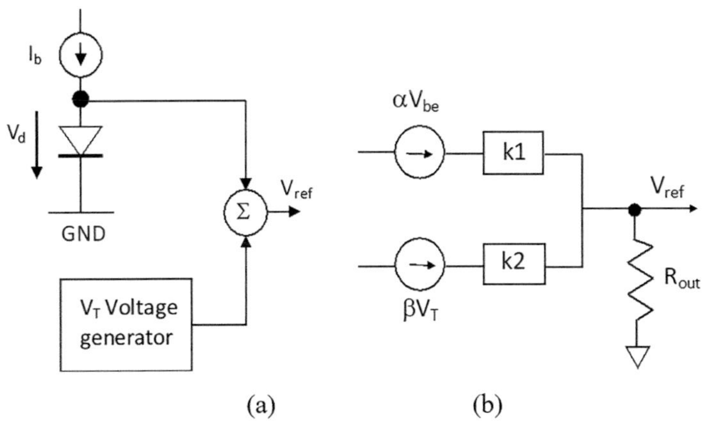

Figure 2.5 Implementation principle of temperature compensation: (a) voltage addition and (b) current addition.

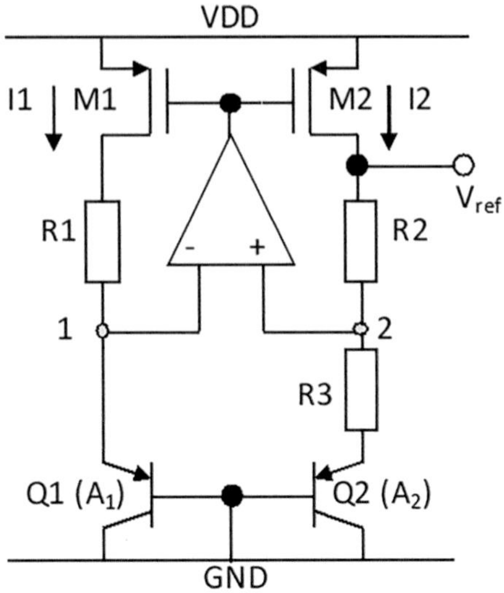

Figure 2.6 Implementation of voltage-based bandgap reference [1].

where $V_{BE}(T)$ is the CTAT term and $V_T(T)$ is the PTAT one. In order to compensate for the CTAT and the PTAT contribution, the factor k shall be chosen accurately. In particular, since it should be equal to the ratio between 2.2 mV/0.086 mV ≈ 25.6.

One possible implementation of the voltage-based principle shown in Figure 2.5(a) is reported in Figure 2.6 [1].

Its topology is somewhat derived from that of Figure 2.3. However, in this case, nodes 1 and 2 are equalized by the opamp closed in a loop, to avoid channel modulation. Moreover, output voltage V_{ref} is built up of two terms:

$$V_{ref} = V_{be1} + (R2/R3)\,\Delta V_{be} \tag{2.6}$$

where V_{be1} is inversely proportional to T, and

$$\Delta V_{be} = V_T \, ln \frac{\left(\frac{W_1}{L_1}\right) A_2}{\left(\frac{W_2}{L_2}\right) A_1} \tag{2.7}$$

where A_1 and A_2 are the areas of the bipolar devices. The first term of eqn (2.6) is CTAT and the second PTAT. Thus, the coefficient k of eqn (2.5) must be equal to

$$k = \frac{R_2}{R_3} V_T \, In \frac{\left(\frac{W_1}{L_1}\right) A_2}{\left(\frac{W_2}{L_2}\right) A_1} = 25.6 \tag{2.8}$$

to compensate temperature effect at first order. All resistors are manufactured using the same kind of material. Base voltage can be chosen conveniently to allow correct common-mode input operation for the opamp. A common choice is to tie bipolar bases to ground and choose an opamp topology whose input stage works at one V_{be} as common mode.

With k factor set to the value of 25.6 and since V_T (at room temperature) is 25.8 mV, the typical value of V_{ref} is somewhat higher than 1.3 V (quite similar to silicon energy gap, from this the name "bandgap reference"), so that it cannot accommodate lower voltages. To overcome this limit, one possible solution is the implementation of the current-mode principle of Figure 2.5(b), as shown in Figure 2.7 [6]. The current-mode approach has become more popular since the scaling-down of technology; it is suitable for low supply voltages, down to 1 V and beyond.

Even in this case, the working principle is the addition of a PTAT current (I_1) to a CTAT current (I_2) contributions fed to an output resistor R_3. The output voltage reference is given by

$$V_{ref} = \frac{R_3}{R_2} V_{be1} + \frac{R_3}{R_1} V_T \ln(n) \tag{2.9}$$

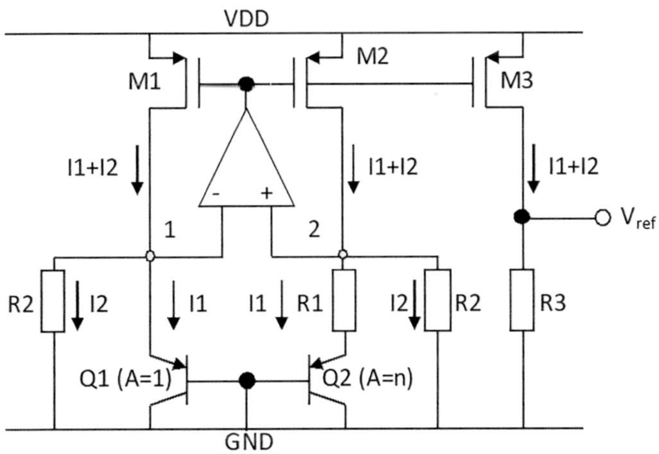

Figure 2.7 Implementation of the bandgap reference based on the current-mode approach [1].

where n is the bipolar area ratio. Also, in this case, it is possible to define a suitable factor k for first-order compensation of temperature. Unlike the voltage-based reference, this circuit can easily implement an attenuation (sub-1 V bandgap reference) or gain of the bandgap voltage by scaling the value of resistor R_3.

The above examples concern only voltage references employing pnp bipolar transistor. A CMOS counterpart can be found, for example, in [7]. This bandgap reference is based on the so-called beta-multiplier principle, where the size of M2 is made larger than that of M1 (its beta is larger than that of M1), so that the difference in the V_{gs} of M1 and M2 is dropped across R. R is an n+ resistor, with a positive temperature coefficient. When the temperature increases, the voltage drop across R increases, too, because of the positive temperature coefficient of the resistor R, and compensates for the decrease of the V_{gs} of M2 due to the negative temperature coefficient of its threshold voltage. Other MOS-based voltage references, which do not make use of temperature-sensitive resistors, generate a PTAT and CTAT contribution (voltages/currents), add them, and send to an output stage [8], [9].

The examples described compensate for reference voltage shift only at first order in a limited temperature range. However, the V_{be} of a BJT is a complex function of temperature, containing many higher-order terms besides the first term. There is a variety of compensation methods in literatures and several different topologies to obtain reference voltages accurate at first and higher orders with temperature (often called curvature compensation), but they are outside the scope of this book and can be found for example, in [2].

2.3 Rad-hard Voltage References

Besides temperature (and supply voltage), radiations add another cause of variability in voltage references. While single-events can cause transient glitches at reference nodes (SET or ASET), total dose can degrade the programmed voltage during component lifetime.

2.3.1 TID counteraction

On the TID side, the variation of the BGR output voltage due to the dose derives from decreased transconductance, elevated subthreshold leakage, and variation in threshold voltage. Several tests have been done in the past on commercial components, showing to what extent these changes can affect device performance [10]–[12]. In particular, the output voltage reference can vary by some percentage point even in the presence of a moderate dose (<100 krad). Since the observed degradation is not negligible, researchers and designers began to study solutions to this issue [13]–[32]. To overcome the total dose effect, starting from the conventional voltage references shown in previous sections, different techniques have been applied at circuit, layout, and process levels to make them resilient to dose. For conventional applications, rad-hard voltage references can be classified based on the generating element. Some significant examples will be described in the following, even if what is shown does not pretend to cover all the solutions proposed. We will see that the most sensitive element to dose is the bipolar transistor used in V_{be}-based topologies, while in MOS-based topologies, obviously, an unavoidable change in threshold V_{th} due to radiation harms the voltage references' performance. An interesting and didactic discussion can be found in [28], where the impact of total ionizing dose effects on the performance of CMOS voltage reference circuits on four circuits with different topologies implemented with a standard 130 nm CMOS process and without any radiation-hardening provisions.

Parasitic diode-based voltage references

The topology that makes use of the thermal voltage V_T to design a voltage generator suffers a certain level of degradation when exposed to radiation doses. The more sensitive device in the scheme of Figure 2.7 (conventional current-mode voltage reference) is the pnp diode-connected BJT. This is an intrinsic effect of a BJT that cannot be avoided [13]. Indeed, for example, in a pnp bipolar transistor, the collector current operating in the forward-active region consists of holes that are injected from the emitter-base junction to the base and then swept through the depletion layer of the collector. In Figure 2.8,

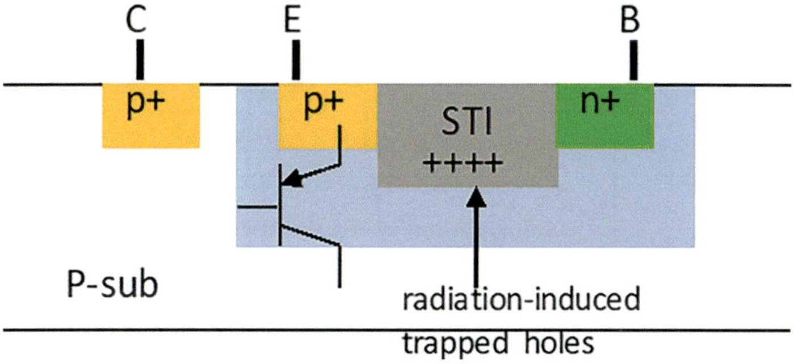

Figure 2.8 Simplified cross-section of substrate-pnp transistor.

the cross-section of a substrate-pnp bipolar transistor, which is available in a standard CMOS technology, is shown. Irradiation induces holes trapping in the body of the field oxide near SiO_2-Si interface. This, in turn manifest in a loss of carriers that recombines in the emitter-base depletion layer or in the base, thus increasing base leakage current, I_B. Since this effect does not significantly change the collector current [14], the BJT current gain $h_{fe} = I_C/I_B$ decreases.

As for radiation with high-energy, BJT exhibits a second effect, called displacement damage (DD), which causes an interaction of radiation with the crystal lattice and produces defects, such as a vacancy where the atoms have been. The macroscopic effect on bipolar behavior is that there is an increase in base current and a degradation in gain.

Further dose effects on the other elements of the voltage reference, such as MOS leakage already described in Chapter 1, shall be considered less relevant [15].

The effect of total dose on bandgap voltage reference with current-based topology could be significant and can be modelled as in Figure 2.9.

The output voltage is no longer represented as in an un-irradiated situation by

$$V_{ref} = \frac{R_3}{R_2} V_{be,Q1} + \frac{R_2}{R_1}\left(V_{be,Q1} - V_{be,Q2}\right) = R_3\left(\frac{V_T In(n)}{R_1} + \frac{V_{be,Q1}}{R_2}\right) \quad (2.10)$$

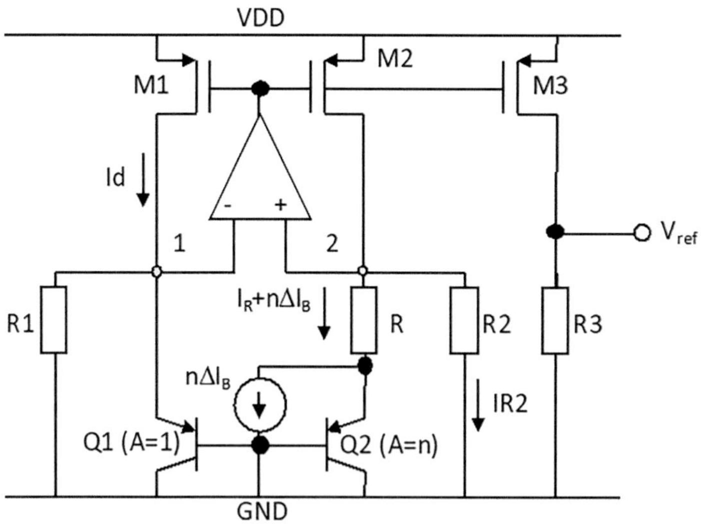

Figure 2.9 Modeled TID effects in the current-based bandgap reference as base leakage currents [15].

While the collector currents stay constant, the emitter current of Q2 (which is n-times bigger than Q1) I_R, must increase because of an increase in its base current. Being $I_R = \Delta V_{be,Q1\text{-}2}/R_1$, while $V_{be,Q1}$ remains unchanged, $V_{be,Q2}$ must decrease to compensate for base leakage current. If $V_{be,Q2}$ diminishes also its collector current, I_{C2}, diminishes, as well, towards a new stable state. If we define $I_{B,Q2} = I_{BQ2,0} + \Delta I_{B,Q2}$, and $I_{C,Q2} = I_{CQ2,0} - \Delta I_{C,Q2}$, where $I_{BQ2,0}$ and $I_{CQ2,0}$ are the pre-irradiated base and collector carriers, $\Delta I_{B,Q2}$, is the increment in base current due to radiation, and $\Delta I_{C,Q2}$ is the collector current decrement, we obtain

$$V_{ref} = R_3\left(I_{CQ2,0} - \Delta I_{C,Q2} + I_{BQ2,0} - \Delta I_{B,Q2}\right) + \frac{R_3}{R_2} V_{be,Q1} \tag{2.11}$$

From eqn (2.11), it can be argued that at the beginning of irradiation, I_{B2} current starts to increase, then after $V_{be,Q2}$ dropping collector current tends to compensate for the further increase of $I_{B,Q2}$ with radiation. Under dose in eqn (2.10), the first term (CTAT) does not change, while the second term (PTAT) increases because of $V_{be,Q2}$ decreases. The net effect is an increase in the voltage reference output voltage. This behavior has been clearly observed in the radiation campaign reported in [15]. While $V_{be,Q1}$ has only a small shift due to temperature change, $V_{be,Q2}$ significantly drops. This reflects immediately on V_{ref}, causing a variation in the reference voltage higher than 15% when exposed to radiation > 4.5 MGray (1 Gray = 100 rad).

A couple of solutions for the BJT leakage issue have been proposed. The first one [15] addresses a very high level of TID typical of high-energy experiments in physics, while the second [16] exhibits good radiation tolerance up to 0.4 MGy.

The first possible solution has been proposed by [15]. It is called dynamic base leakage compensation (DBLC) and provides the base current needed by a BJT closed in a diode. It relies on a feedback compensation circuit, which makes all the base leakage currents induced by radiation flow only through the compensation network, while currents in the upper MOS remain constant. The principle is shown in Figure 2.10.

The proposed principle works as follows. Opamp OTA1 and OTA2 make nodes $V_{bl} = V_{br} = V_{bref}$. This, in turn, makes $I_{R1} = I_{R2} = I_{R3}$. Since M1, M2, and M3 have the same size, their drain current is almost the same and equal to I_{R3}. From Kirchhoff's law, we also have that $I_{R1} = I_{D,M1} = I_{D,M6} + I_{B,Q1} = I_{D,M6} + I_a$. From this derives that $I_{B,Q1} = I_a = I_{B,Q2}$. If during irradiation the base current increases, as explained above, the feedback loop compensates dynamically through I_a. Additional auxiliary circuitry, not shown, helps to reinforce the equalization of currents in upper MOS transistors.

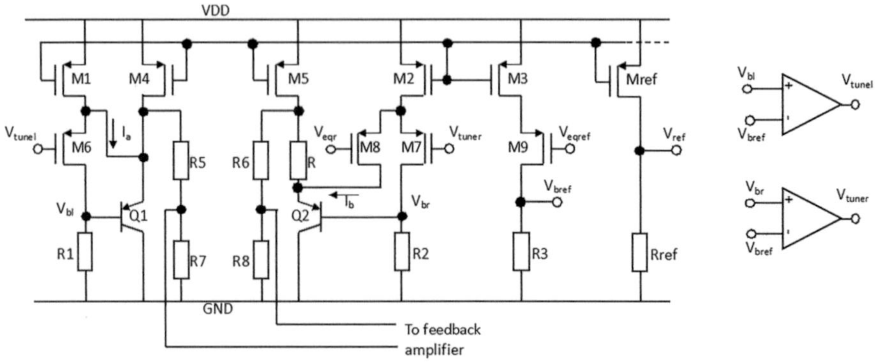

Figure 2.10 Simplified scheme of the voltage reference with DBLC technique [15].

The modified voltage reference with DBLC principle has been integrated in a standard 130 μm CMOS process, and experimental results confirm the effectiveness of the DBLC technique under a real-time ^{60}Co gamma irradiation assessment. The output voltage of the bandgap reference is designed to be 600 mV. Measurements showed that, thanks to the dynamic base leakage compensation technique, both $V_{be,Q2}$ and $V_{be,Q1}$ are immune to the total ionizing dose. This is reflected in the behavior of the output voltage reference. Figure 2.11 shows the output voltages of the radiation-hardened

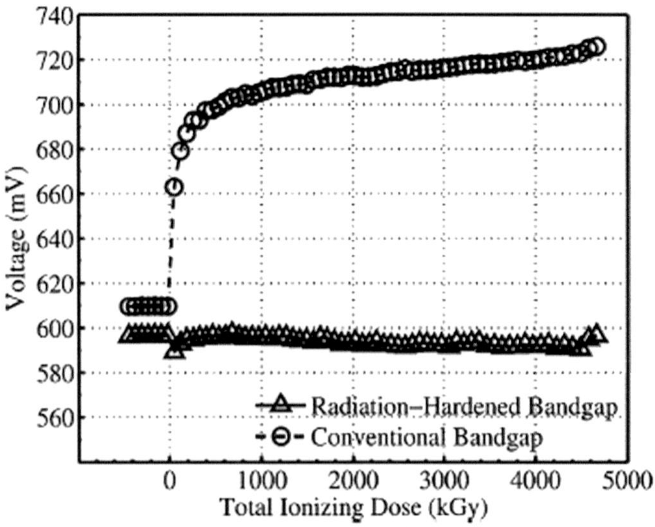

Figure 2.11 Measured output voltages of the radiation-hardened bandgap reference and the conventional bandgap reference [15].

bandgap reference and the conventional bandgap reference of one measured sample. A variation of only 1% is found up to 4.5 MGray, while the uncompensated features 15% voltage shift.

The variation measured on five samples is <3% only. Behavior in temperature does not differ from that of conventional architectures.

A second solution to the issue on BJT considers as key point the use of an alternative element in place of the pnp-diode: indeed, the conventional diodes are replaced with a so-called dynamic threshold MOS transistor (DTMOST) with ELT layout [16]. These devices are inherently rad-hard resilient and have a diode-like I–V. It is a de facto PMOS with gate, drain, and substrate connected and operates as a diode with low-bandgap voltage [17]. From a physical point of view, its insensitivity to radiation arises from the absence of any thick oxide near the p-diffusion region (Figure 2.12). This differs from a conventional diode, where such oxide layers are present and can trap radiation-induced holes. From a top view, the DTMOST shape looks like an ELT (or annular transistor).

The starting point for the proposed voltage reference, designed in a 130 nm standard CMOS technology, is a conventional voltage-based bandgap reference. However, the pnp bipolar transistors have been replaced by DTMOST (Figure 2.13), having an exponential behavior. Voltage V_{ref} is expressed by

$$V_{ref} = V_1 + \frac{R_1}{R_2} V_T \ln(n) \tag{2.12}$$

where n is the ratio between the areas of the diode devices and V_1 is the voltage of the negative input of the opamp. When a DTMOST in a weak-inversion

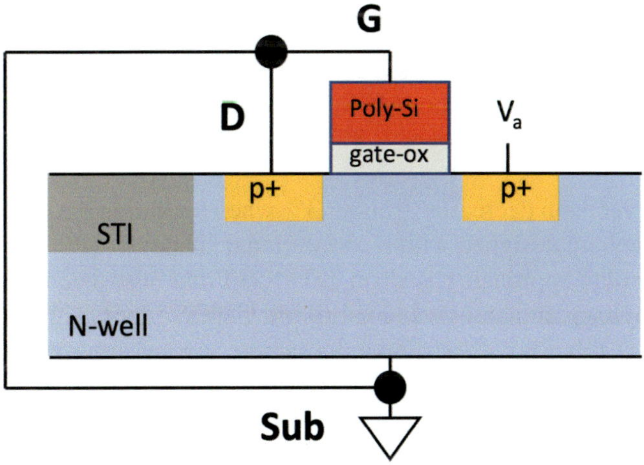

Figure 2.12 Cross-section of a DTMOST [16].

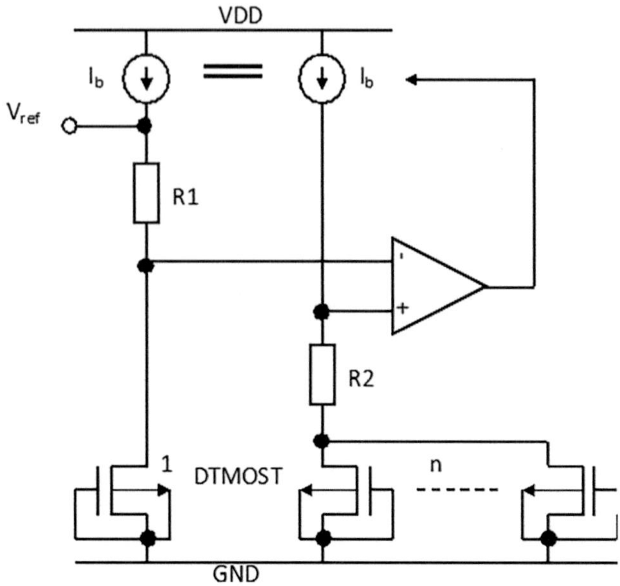

Figure 2.13 Scheme of the bandgap voltage reference circuit with DTMOST diodes [16].

region is used in place of a conventional pnp transistor, we make it operate in the region of exponential behavior, starting from a junction forward biasing as low as 0.1–0.2 V (instead of 0.5–0.6 V as in a normal diode). In this way, the output reference voltage of the present circuit will be much lower than that for the typical CMOS bandgap voltage reference. The proposed circuit has been designed featuring V_{ref} around 410 mV with an untrimmed temperature gradient (at constant current) of about 0.8 mV/°C.

Measurements on integrated prototypes confirmed that if the ELT shape is employed also for the other MOS, the bandgap and the reference voltage remain stable with accuracy < 0.8% up to 44 Mrad (X-ray), as shown in Figure 2.14.

The precision decreases to about 3% when including the contribution of MOS, if they are designed with a conventional shape.

A similar approach based on DTMOST has also been adopted in a second topology implemented in a 180 nm CMOS process [18]. In addition to [16], this topology also allows correcting second-order effects induced by radiation, such as opamp offset, current mirror mismatch offset, and DTMOST offset. The proposed scheme is shown in Figure 2.15, having as a starting point again the topology of Figure 2.13.

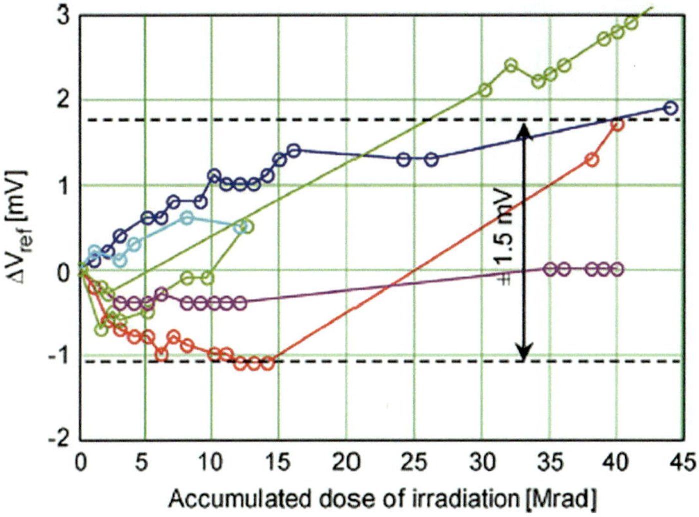

Figure 2.14 Measured shift of the reference voltage during irradiation for 6 prototypes [16].

Two main concepts are introduced here: chopping and an architectural change in the upper current mirror. As for the former, it is known that the technique of chopper stabilization is well-suited to a DC circuit like a voltage reference and allows for dynamically reducing the offset associated with the opamp. Secondly, one PMOS current source is eliminated, while polysilicon resistors with values R and (R/n) are used to scale the bias currents of the DTMOSTs by a factor of n. In this way, by eliminating the current mirror, one of the dominant sources of radiation-induced mismatch is removed, and the two DTMOST can be made equal. Since polysilicon resistors are extremely resistant to TID-induced mismatch, any mismatch associated with scaled currents is cancelled. Measurement was done by irradiating with X-rays some prototypes fabricated in a 180 nm standard CMOS technology. The initial voltage of static (chop clocking grounded) and chopped voltage references were 489 mv and 527 mV, respectively. Measured shifts in reference voltage after irradiation are shown in Figure 2.16.

The static reference showed a maximum shift of about 176 mV and a shift of 32 mV at 1 Mrad (36.0% and 6.6%, respectively). Instead, the chopped reference showed a maximum shift of 6.0 mV and a shift of 4.6 mV at 1 Mrad (1.1% and 0.8%, respectively). This was a 96.6% reduction in overall shift.

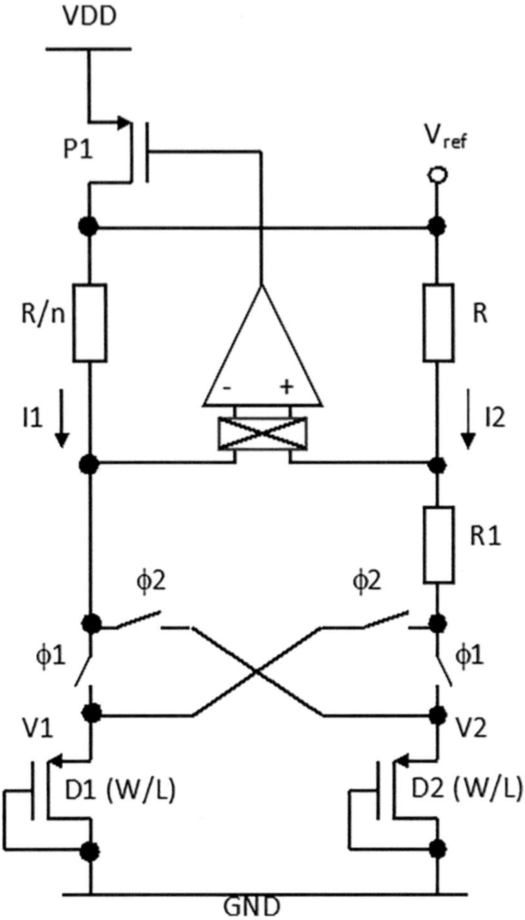

Figure 2.15 Modified radiation-hardened voltage reference design [18].

A further bandgap topology, which makes use of a diode-connected transistor, is that one in [19]. It is derived from conventional current-mode architecture, where a PTAT and a CTAT current are added [6]. The pnp bipolar transistor, however, is replaced with a diode build-up using a PMOS transistor with an ELT shape and where drain and gate are connected to ground, while the source represents the diode anode and the bulk (n-well) the cathode (Figure 2.17). The active area is surrounded by gate oxide instead of STI; indeed, the inner p+ diffusion together with the n-well can be considered like a diode with the STI oxide moved away [20].

Thanks to the ELT shape, the choice of W and L of MOS transistors greater than a certain minimum value, and other circuital precautions, the voltage reference is stable within 10% for a dose up to 800 Mrad.

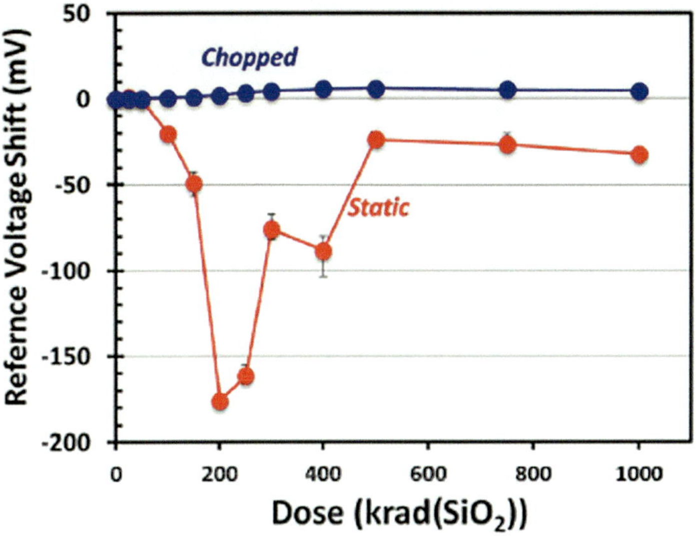

Figure 2.16 Measured shifts in reference voltage for the chopped and the static references on integrated prototypes [18].

We want to point out another interesting publication [21], which starts from [16] and [19] and proposes a comparison of post-layout simulations of voltage reference circuits and PTAT voltage sensor implemented using either NMOS bulk diode, as in [19], or DTMOS, as in [16], in a 130 nm technology. The impact on temperature performance of such devices, used as a replacement for a conventional p-n junction (or a diode-connected bipolar) is analyzed.

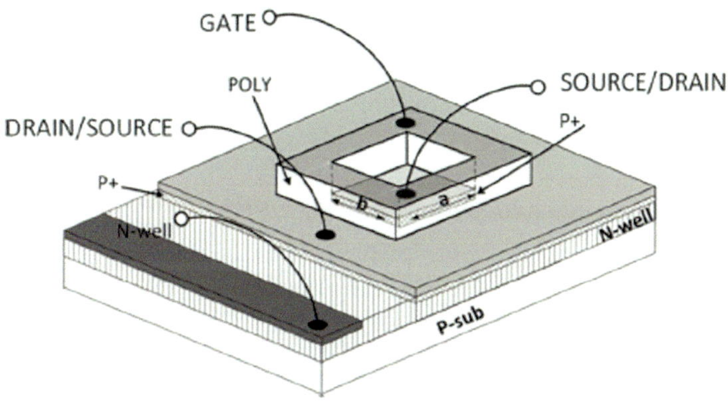

Figure 2.17 Enclosed-layout diode based on a PMOS [19].

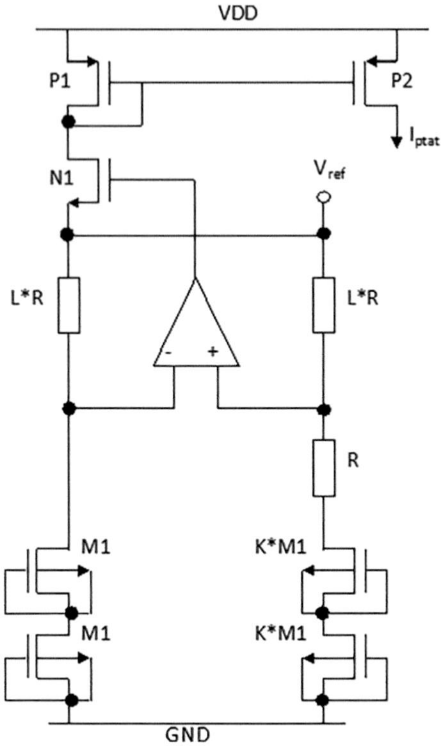

Figure 2.18 Scheme of the voltage-based voltage reference circuit in [22].

Starting from proposals [18] and [19], also [22] presents a 5 V voltage reference which replaces p-n diode or bipolar and operates with two stacked DTMOST (allowed by high supply voltage) in a 180 nm CMOS process (Figure 2.18). Indeed, as said, the p-n junction diodes are the dominating factor in radiation degradation in bandgap reference circuits.

The stacking design has been chosen to allow the DTMOS bandgap reference design to utilize the conventional bandgap reference circuit structure, such as with a p-n diode or bipolar (remember that DTMOS has a lower bandgap voltage, as we have underlined above).

The equation that regulates the operation of this voltage reference is still

$$V_{ref} = V_1 + V_T L \ln(K) \tag{2.13}$$

where V_1 is the voltage at the input of the opamp (and at the drain of M1), K is the ratio between DTMOS areas, and L is the ratio between resistors, and is chosen to cancel the CTAT portion of the reference voltage. The first term in eqn (2.13) is CTAT, while the second is PTAT. For DTMOS, the CTAT term, V_1, has a logarithmic relationship with the current through it, exactly as in a conventional diode, but with a lower bandgap potential [23]. However, thanks to the series arrangements, V_{ref} can be approximately 1 V. This project choice also increases the voltage drop across the resistor R, thus reducing the effect of offset voltages induced elsewhere in the bandgap circuit.

Experimental results on integrated prototypes with DTMOS having annular gate showed a 21.2 mV variation over the military temperature range (–55°C to 125°C), which means 2%, while V_{ref} varies of about 5 mV up to 300 krad dose (0.45%).

All MOSFET-based voltage references

A second architectural approach exploits the characteristics of MOS-only transistors to compensate for temperature and radiation effects in a voltage reference.

The solution proposed in [24] combines linearly two different NMOS threshold voltages to generate a CTAT current and compensate radiation induced effects. Then, the CTAT and the PTAT voltages are further combined to allow the compensation of temperature-induced variation. It is a current-mode topology, whose principle is shown in Figure 2.19.

The first core sub-circuit is the current generator, which generates a CTAT current starting from the threshold voltage difference (M1–M7). Each transistor works in the strong-inversion region. The OTA keeps $V_{ds1} = V_{gs2} = V_{ds2}$. From circuit inspection, it can be derived that

$$I_1 = \frac{V_{gs1} - V_{gs2}}{R_1} = \frac{\Delta V_{th}}{R_1} \tag{2.14}$$

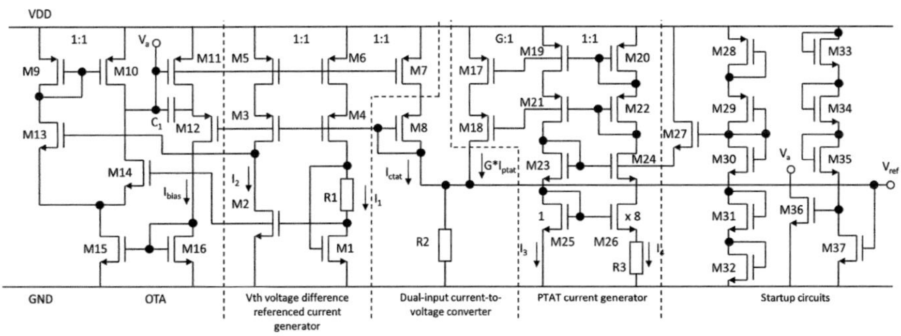

Figure 2.19 Scheme of the voltage reference circuit in [24].

assuming that the channel length effect dependencies of M1 and M2 are equal (at first approximation). Transistors M1 and M2 are implemented with different gate lengths (L) to achieve different threshold voltages, V_{th}. ΔV_{th} is the difference between threshold voltages and is proportional to the Fermi potential, which decreases with temperature [25]. Since TID effects can be assumed independent of MOS size [26], the two threshold voltages exhibit the same radiation degradation and cancel out from eqn (2.14).

The second key sub-circuit is the PTAT current generator. Transistor M25–M26 operates in the subthreshold region (favorable in terms of TID, as we will see later). By making the M26 the combination of eight instances of M25 (so they have the same L and the same threshold), the PTAT current can be written as

$$I_2 = \frac{nV_T}{R_3} \ln(8) \tag{2.15}$$

where V_T is the usual thermal voltage and "n" is the subthreshold slope factor for MOS [1]. Also, I_2 can be considered almost independent of dose.

Finally, MOS M7–M8 and M17–M18 combine CTAT (I_1) and PTAT (I_2) currents with a gain G

$$V_{ref} = R_2\left(I_{CTAT} + GI_{PTAT}\right) = \frac{R_2}{R_1}\Delta V_{th} + \frac{GR_2}{R_3} nV_T ln(8) \tag{2.16}$$

By choosing a scaling factor, it is possible to minimize the temperature dependence of V_{ref} at first order. Moreover, more important I_{CTAT} (I_1) and I_{PTAT} (I_2) are almost independent of radiation, making V_{ref} very stable also at high TID. Measurements on integrated prototypes implemented in 130 nm CMOS technology with transistors designed with conventional shapes (not ELT), showed that the variation of V_{ref} (fixed around 718 mV) is lower than 1% up to 145 krad.

One more radiation-hardened voltage reference with MOS operating in the subthreshold region is proposed in [27]. It is based on the current summing approach and features a low-voltage supply up to 0.75 V. Subthreshold MOS operating at a low supply voltage shows two opposite behaviors under TID. A lower electric field across oxides results in the separation of electrons and holes and increases recombination. However, the drain current is exponentially dependent on threshold V_{th}, and thus on irradiation if V_{th} changes:

$$I_D \,\alpha\, MV_T^2 \exp\left(\frac{V_{GS} - V_{th}}{nV_T}\right) = \frac{R_2}{R_1}\Delta V_{th} \tag{2.17}$$

where M includes the MOS size, carrier mobility, and oxide capacitances per unit area and slope factor. The proposed voltage reference is shown in Figure 2.20. It consists of three sections: (a) core circuit that provides temperature correction, (b) PTAT bias current generator, and (c) output reference voltage generator.

This scheme realizes a linear approximation through the addition of PTAT and CTAT currents, while higher-order temperature dependency (non-linear curvature correction) is achieved using polysilicon resistors

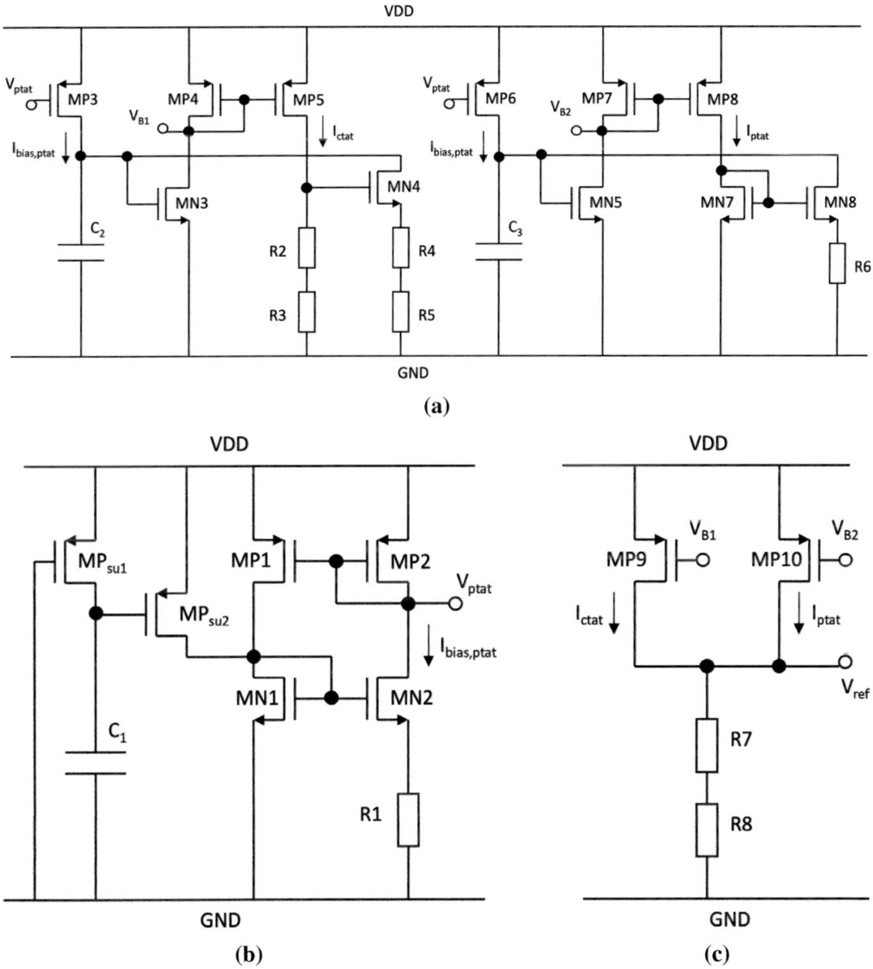

Figure 2.20 Scheme of the voltage reference in [27], operating in the sub-threshold region.

having complementary responses to temperature (such as high-ohmic R_2 and medium-ohmic R_3) and the MOS MN7. The output voltage is expressed by

$$V_{ref} = \left(I_{PTAT} + I_{CTAT}\right)\left(R_7 + R_8\right) \tag{2.18}$$

By circuit inspection, the detailed expression of V_{ref} is given by [27]

$$V_{REF} = \frac{R_{7,8}}{R_{2,3}} V_{GS4} + \frac{R_{7,8} R_{4,5}}{R_1 R_{2,3}} U_T \ln\left(\frac{K_{MN2}}{K_{MN1}}\right) + R_{7,8} K_{MN7} I_0 \exp\left(\frac{V_{GS7} - V_{TH}}{nU_T}\right) \tag{2.19}$$

where U_T is the thermal voltage and K are the transistor's size aspect ratio.

By adjusting the resistors with different temperature coefficients, it is possible to compensate linearly and at higher order temperature dependencies.

Since this circuit does not use bipolar, it instead utilizes MOS operating in the sub-threshold region, with a maximum stack of two MOS. The radiation induced variations in the two loops (PTAT and CTAT) approximately compensate each other, and the output voltage reference exhibits good resiliency to TID. The well-known radiation hardening by design (RHBD) technique, such as annular transistors or ELT, has not been utilized since it's important maintain good performance, when MOS transistors are biased in the sub-threshold regime (the drain-current is exponentially dependent on threshold voltage).

Measurements on integrated samples in a standard 180 nm CMOS technology showed that after trimming, it achieves 15 ppm/°C over a –40° to 130°C temperature range. And, most importantly, after a multi-step irradiation up to 4 Mrad, the reference shows –0.7% to 1.2% maximum deviation from the original output voltage V_{ref}, as shown in Figure 2.21.

2.3.2 ASET mitigation

As concerns the transient phenomena, much attention has been paid to the analysis of SET in BGR circuits and the radiation-hardened-by-design techniques. Single-events usually arise from the impact of an energetic heavy ion or proton, possibly leading to single event upsets (SEU) or single event latch-up (SEL). SEU are mainly pertinent to digital circuits where they change logic states (e.g., flip-flops) by dumping relatively large charges on critical (bit flipping). In an analog circuit, such as a voltage reference, a transient pulse can be observed, but after a certain amount of time, it subsides

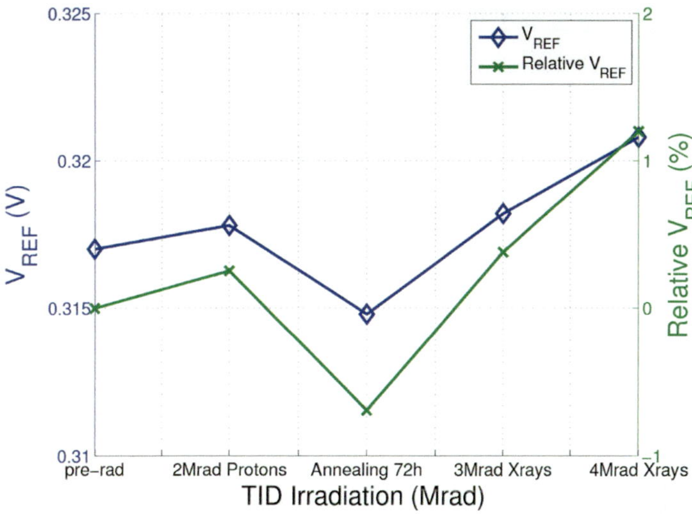

Figure 2.21 Voltage reference variation after multi-step irradiation up to 4 Mrad for the circuit in [27].

(analog single event transient (ASET)). These phenomena can bring severe issues to several analog circuits, for example, in data converters [33], where a bit-flip in the internal logic can completely distort the results of one conversion. However, also in voltage reference, which has inherently a single stable state and where the accuracy of the generated voltage is a crucial parameter, single-events can be a concern.

Different RHBD techniques have been proposed, mainly including the multi-modular redundancy technique, isolation technique, and RC filter technique [34]–[37]. For example, the ability of SET tolerance has been improved by employing the dual modular redundancy (DMR) technique in the BGR circuit [34]. In [35], it indicates that an obvious suppression of SET is achieved by introducing an inverter and a transistor transmission gate in BGR. However, the width of SET is still unchanged. The solution proposed in [36] combined a nonlinear transform module with an RC filter module (with a 0.5 pF capacitor) to realize the SET mitigation. The results show that the proposed modules can be used to achieve effective SET filtering. An analysis of the reported technique (including DMR, isolation technique, and RC filter technique) shows that a good suppression of the SET pulse is achieved. However, it should be noticed that the reported RHBD technique had increased significantly the layout area due to the redundancy approach or the addition of capacitors in hardened BGR circuits. What is more, resistors and capacitors used to filter such a long-lasting perturbation tend to increase

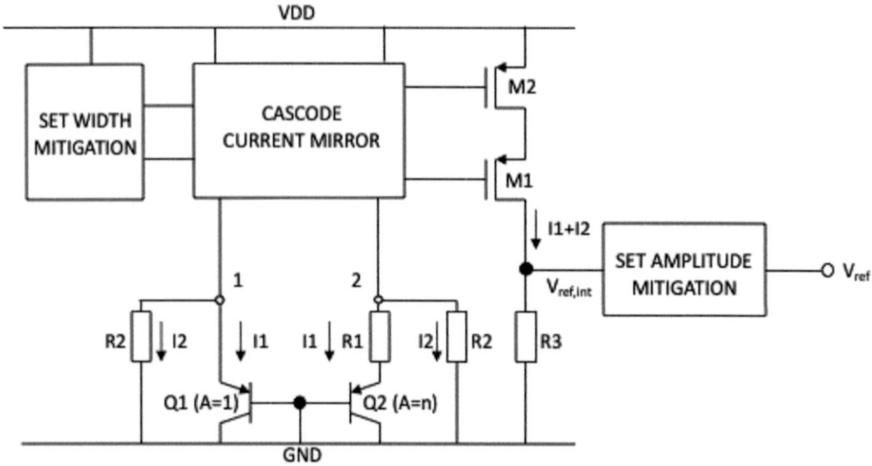

Figure 2.22 Proposed SET-tolerant bandgap design [37].

the system recovery time. A recently presented solution that seems to overcome this limit has been proposed in [37]. Here (Figure 2.22), two additional radiation-hardened peripheral circuits, consisting of a SET width mitigation module and a SET amplitude mitigation module, are added to a typical bandgap reference circuit (such as that in Figure 2.7, with the opamp replaced by a cascode current mirror). The SET width module compensates heavy ions strikes in the cascode section by injecting a suitable number of opposite charges. The SET amplitude module disconnects the internal output of the bandgap, $V_{ref,int}$, when a SET is detected. The effectiveness of this technique is demonstrated with post-layout simulation in 55-nm CMOS technology.

In [38], another interesting investigation of space radiation induced ASETs due to heavy ions, when designing reference circuits for space electronics can be found. Three bandgap voltage references are designed with different topologies and technology characteristics, and the goal is to identify a minimum set of design strategies as "rules of thumbs" useful to mitigate ASETs effects. The chosen topology is based on the principle of summing a PTAT voltage and a CTAT voltage to compensate for the temperature effects and achieve a temperature-independent voltage at the output (at first order). But instead of using a feedback loop to equalize emitters voltages of the two bipolar transistors, as in Figure 2.6, a current mirror is used. The two topological variants are shown in Figure 2.23 and Figure 2.24. The former includes a simple current mirror and has been analysed, featuring two different supply voltages: 1.8 V and 3.3 V. The latter operates at 3.3 V supply and comprises a cascode current mirror.

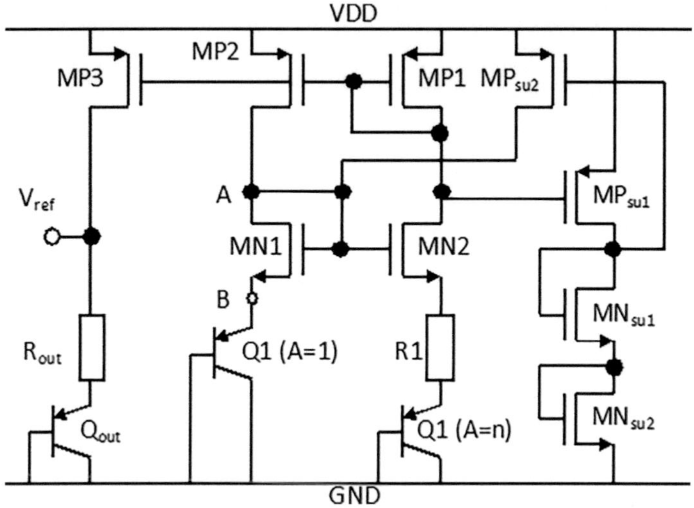

Figure 2.23 Scheme of voltage-based bandgap-reference according to first variant, SCM - Simple Current Mirror (both with 1.8 and 3.3 V voltage supply) [38].

Besides architecture, the two circuits also differ with respect to the oxide thickness of the utilized transistors. Indeed, the used 180 nm standard

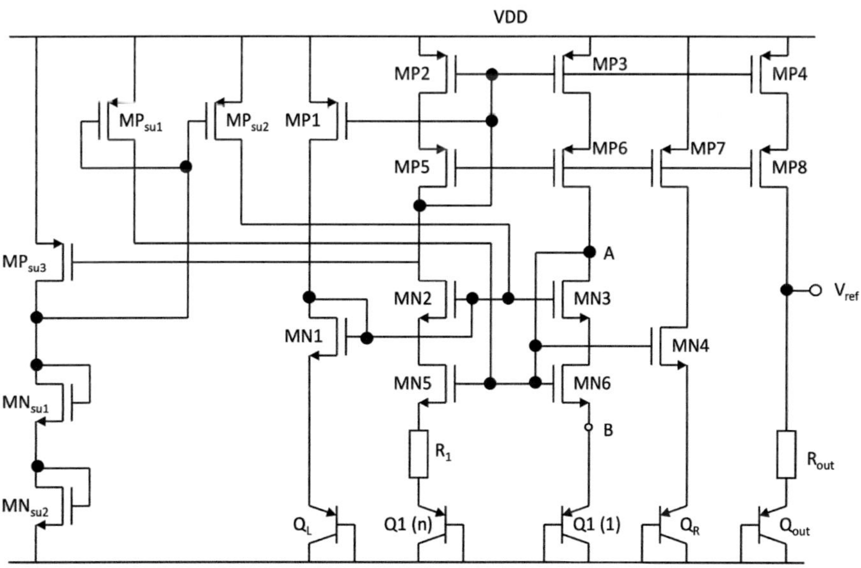

Figure 2.24 Scheme of voltage-based bandgap-reference according to the second variant, CCM - Cascode Current Mirror (3.3 V voltage supply) [38].

CMOS process makes available 1.8 V thin-oxide MOS and 3.3 V thick-oxide MOS.

The reference voltage at the output of both topologies can be expressed by

$$V_{ref} = V_{be,Qout} + \frac{R_{out}}{R_1} V_T \ln(n) \tag{2.20}$$

where n is the bipolar areas ratio. The first term is a CTAT component, and the second term is a PTAT component. A low variation with temperature can be achieved by properly tuning the ratios of n, R_{out} and R_1 terms.

In this design, standard RHBD provisions have been adopted in layout, such as guard rings and well-substrate contacts in all circuits to improve the robustness to device latch-up, increased transistor sizes for hardening against ASETs, while edgeless NMOS devices were employed for hardening against total ionization dose (TID) radiation induced leakage.

The reported test indicates that the integrated prototypes have been exposed to a cocktail of heavy ions (Si, Kr, and Xe), and the resulting ASETs were recorded at the output, V_{ref}, of each topology by using a high-performance oscilloscope. The cocktail provides Si ions with a LET(Si) of ~ 6.9 MeV·cm²/mg, Kr ions with a LET(Si) of ~ 36.1 MeV·cm²/mg, and Xe ions with a LET(Si) of ~ 64.7 MeV·cm²/mg. The bandgap references were biased at nominal supply voltages during irradiation. First observed parameter is the cross-section σ in cm², defined as the ratio between the number of observed ASET and the ion fluence Φ ($\sigma = N_{ASET}/\Phi$). The threshold for ASET has been set at 10 mV, above the noise floor. Test indicated that all three different topologies do not exhibit any significant sensitivity to Si ions (LET~ 6.9 MeV·cm²/mg) since such particles were not able to induce detectable transients to any of the bandgaps. For Kr ions, the bandgap of Figure 2.22 at a 1.8 V circuit exhibited a much higher σ, when compared to the same with a 3.3 V supply, and slightly higher σ when compared to the bandgap in Figure 2.23. While using Xe ions, the bandgap of Figure 2.22 at 3.3 V still has the lowest σ, the σ of the same circuit at 1.8 V is between the other two. This anomaly can only be explained by the pulse quenching phenomenon, which has been observed in other circuits and explained in more detail in [36]. As further information, a recorded ASET response to Xe ions, obtained from the circuit in Figure 2.22 with a 1.8 V supply, indicates that the ASET duration is approximately 10 μs and its amplitude is approximately 55 mV.

Finally, Figure 2.25 shows the measured number of ASETs vs ASETs amplitude (left) and duration (right) for all bandgap voltage reference variants (SCM = simple current mirror, Figure 2.23, CCM = cascode current mirror, Figure 2.24) for Kr ions. Same graphics for Xe ions can be found

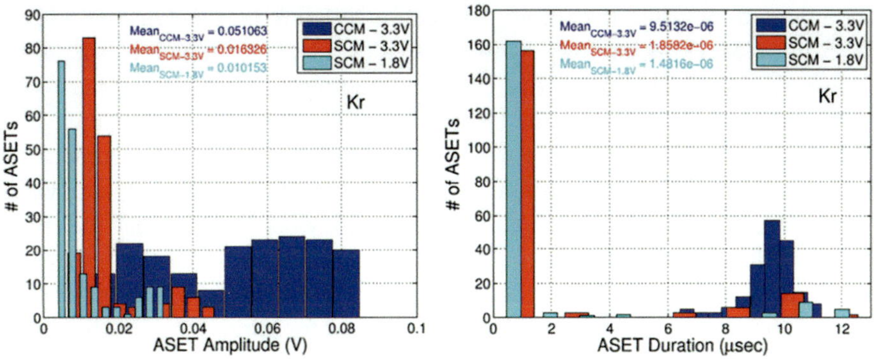

Figure 2.25 Measured number of ASETs vs ASETs amplitude (left) and duration (right) for all bandgap voltage reference variants (SCM = simple current mirror, CCM = cascode current mirror) for Kr ions [38].

in [38]. From the pictures, it can be deduced that both SCM circuits have smaller ASET durations when compared to the CCM bandgap. Moreover, the SCM topologies feature smaller ASETs peak amplitudes when compared to the CCM topology, with the SCM-1.8 V circuit exhibiting the smallest peak amplitudes. This can be explained by the fact that the lower impedance at critical nodes enhances the escape speed of the ionization charge towards the supply or ground. In general, all three circuits exhibit increased ASET amplitudes for Xe ions compared to Kr ones due to their increased LET.

Overall measurements suggested that better ASET performance is achieved by the simple mirror bandgap operating at 3.3 V, because of smaller path resistance to supply rail/ground rail as opposed to the cascode current mirror bandgap at 3.3 V. Moreover, a thicker oxide results in smaller transconductance and thus, smaller transient amplification. Generally, better performance is achieved with thicker oxides, higher supply voltage, and fewer cascode stages. These guidelines apply to all similar analogue and mixed-signal circuits.

2.4 New Materials for Voltage References

So far, we have explored the existing solutions that have been proposed to overcome the sensitivity of voltage references to radiation and which deal with an RHBD approach at circuit and physical layout levels. Besides making a rad-hard design, technology can also be a help, even if we have underlined in the introduction that leveraging the fabrication process only is not always the best solution. For example, the low-voltage (LV) regulator in [39], including three BGRs, exploits the characteristics of the Onsemi I3T80 0.35 um CMOS technology, which provides adequate radiation tolerance. However, the production of this process is going to cease. Two other technologies can help in hitting the target: SOI (Silicon-On-Insulator) and SiGe (Silicon-Germanium). The former is particularly useful for the implementation of all-MOS voltage references, while the second represents a valid replacement for conventional bipolar technology, as we said that the main cause of degradation in such circuits is the pnp (or npn) transistors.

An example of a SOI voltage reference is given in [40]. The SOI process is often used in harsh environment applications, for attractive some features: an extended range of working temperature, low-power, high speed, and lower sensitivity to transient radiation effects compared to other standard CMOS technologies [41]. In particular, in [40] an SOI process of 130 nm is used, having a thin gate oxide (5 nm) and hence good resistance to TID degradation. The MOS in the circuit operates in sub-threshold, exploiting the advantages already underlined above. The scheme of the circuit is in Figure 2.26.

Its working principle is based on V_{gs} compensation, by generating a gate-source voltage difference between a PMOS (M_P) and an NMOS (M_N) biased by a PTAT current I_{REF}. The expression and dependencies of the output voltage from circuit parameters can be found in [40].

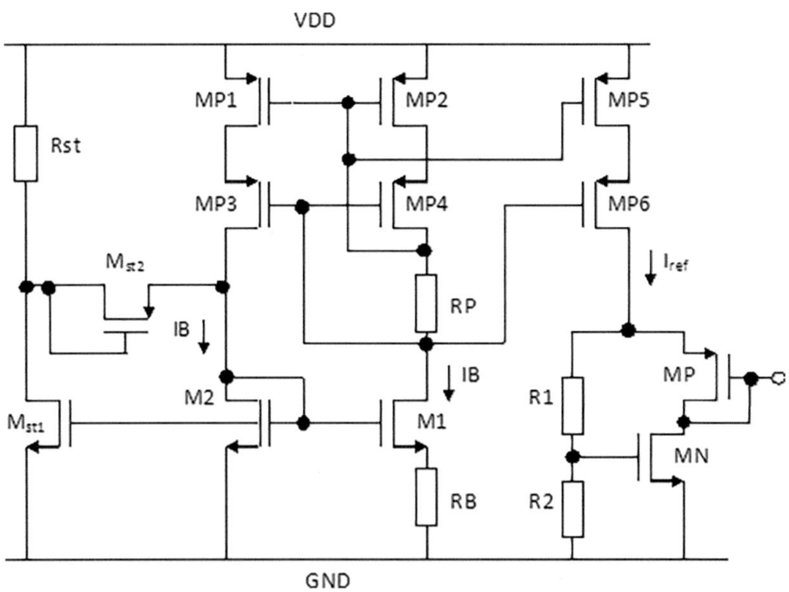

Figure 2.26 Scheme of the SOI voltage reference proposed in [40].

Thanks to SOI characteristics, operation in sub-threshold and layout implemented using specific guidelines to limit TID effects, measurements done with gamma ray (Cobalt source) confirmed the effectiveness of the proposed circuit. The output voltage V_{ref} showed a temperature coefficient of 470 ppm/°C (around 11% over –40–200°C temperature range), while it exhibits a variation of 25 ppm/krad, which means 5% up to 1.5 Mrad.

As concerns SiGe, very interesting works have been presented in [42] and [43]. Indeed, unlike the conventional bipolar process (or the npn/pnp available in a standard CMOS process), the SiGe heterojunction bipolar transistor (SiGe HBT) is intrinsically resilient to the dose, as demonstrated in several studies [17], [44]. To better study the behavior under radiation of SiGe voltage references, two different bandgaps have been implemented in a BiCMOS 0.35 um technology. The first one features only first-order compensation with any further curvature compensation (Figure 2.27).

The output voltage V_{ref} is given by the summation of two voltages, the first one being conversely proportional to temperature, the second one being proportional to temperature, since derived from a PTAT current:

$$V_{ref} = V_{be3} + k\frac{\Delta V_{be1,2}}{R_1}R_2 \tag{2.21}$$

where k is the upper mirror ratio. This scheme compensates only first-order temperature dependency. Higher-order terms in the function of temperature can

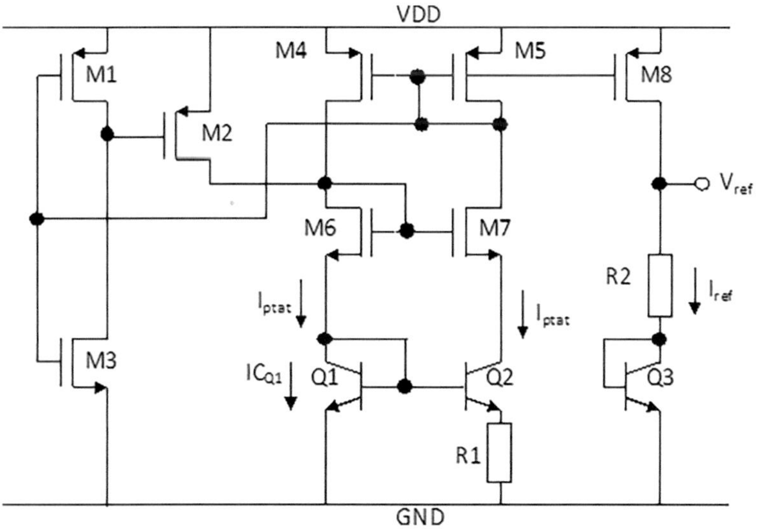

Figure 2.27 Schematic of first-order temp-compensated voltage reference [43].

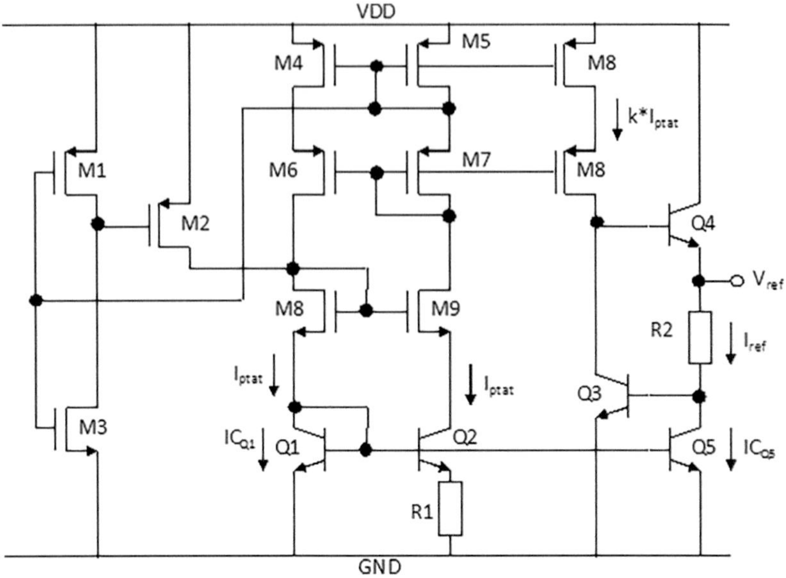

Figure 2.28 Schematic of curvature compensated voltage reference [43].

cause output voltage shift. A second bandgap topology, shown in Figure 2.28, has been implemented, which features curvature compensation.

In this scheme, the compensation is achieved by leveraging the temperature behavior of bipolar Q_3. In eqn (2.21) a third term is added, $k_o R_2(\Delta V_{be1,2}/R_1\beta_{Q3})$, where k_o is the amplification factor of the last stage, and β_{Q3} is the gain factor of Q_3. Since this gain has an exponential temperature dependence, we achieve an approximate compensation of higher-order terms.

The two bandgaps have been integrated and V_{ref} measured under radiation at room temperature with 10 keV X-rays up to TID of 1.080 Mrad (SiO_2) and 5.4 Mrad (SiO_2). Figure 2.29 illustrates the obtained results at 1.080 Mrad.

The change in the output voltage of the two circuits is almost negligible after X-ray exposure: the output voltage shift is 2 mV (0.16%) for uncompensated and 3 mV (0.25%) for compensated. At 5.4 Mrad (SiO_2), the degradation is more evident (1.2% and 2.4%, respectively), especially in compensated bandgap. This is due to the fact that any change in the base current after irradiation results in almost a 3× shift in the magnitude of the output voltage in the compensated BGR than in the first-order. A further irradiation with 63.3 MeV at 5.4 Mrad (SiO_2) shows less degradation than X-ray irradiation.

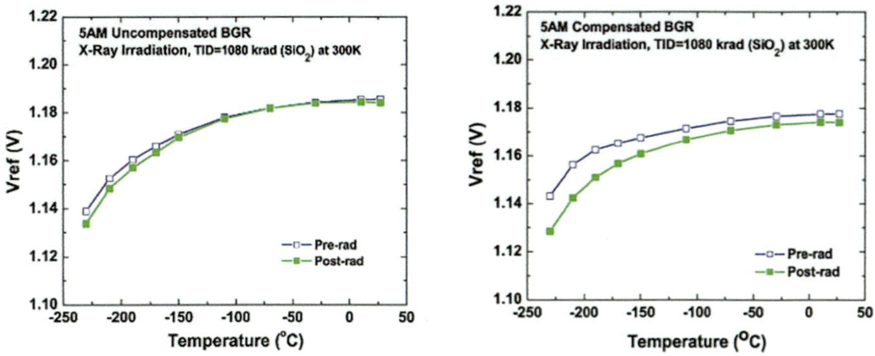

Figure 2.29 Measured reference voltage of uncompensated (left) and compensated (right) bandgap before and after X-ray irradiation with TID = 1.080 Mrad (SiO_2) [43].

Another proposal on SiGe-based references is given in [45]. Here, the design has derived from a voltage reference topology quite like that of the first-order in Figure 2.27. Three variants are studied: with npn SiGe HBT as V_{be} generators, with IM (inverse-mode) SiGe HBT, and a PIN diode, made available by the 90 nm SiGe BiCMOS process. Inverse-mode means that devices operate with the collector and emitter terminals electrically swapped (Figure 2.30). The PIN diodes are made of a p+/n junction; the fabrication steps are like the extrinsic base-collector junction of the SiGe HBT. SET and

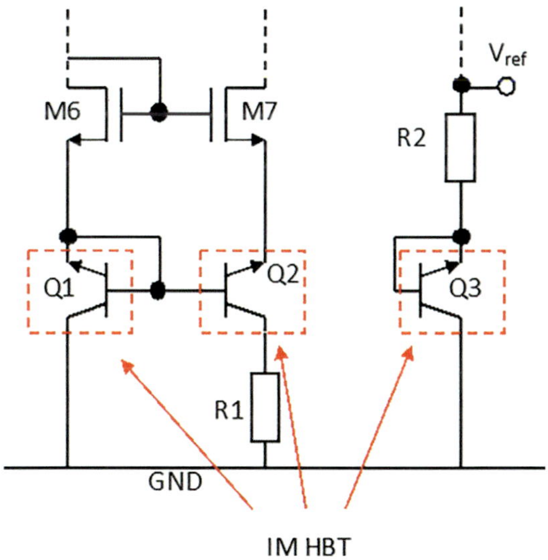

Figure 2.30 Detail of the inverse-mode SiGe HBT used instead of a conventional SiGe HBT.

TID tests have been performed on the three circuits. The presented results demonstrate that SiGe-based voltage reference circuits are sensitive to SET, but also show minimal degradation to TID exposure. While the IM-based bandgap is less sensitive to SET than the conventional scheme, the proposed PIN diode approach demonstrates a higher degree of immunity to SETs than that which uses IM SiGe HBTs.

As a conclusion for the paragraph, we would like to mention new materials which are the new frontiers for high-temperature. Indeed, extreme temperature is one more factor besides radiation that results in critical conditions in a harsh environment. For example, in [46], a bandgap voltage reference for high-temperatures using SiC (silicon carbide) was realized, which still has some lack of stability in the fabrication process and is less accurate than silicon bandgaps, but it can operate up to 540°C, as demonstrated in measurements, a temperature that can be faced in space missions.

2.5 3.3 V 180-nm CMOS Rad-hard Bandgaps

Starting from the consideration that the simpler a solution is, the less likely it will be to degrade under radiation, the authors developed and verified six different bandgaps. The technology used was a 130-nm CMOS. The target performances of the prototypes are Total Dose immunity up to 300 krad (Si) and resiliency to latch-up up to 60 MeV*mg/cm^2.

The conceptual schematic of the bandgap is shown in Figure 2.31. It is based on the so-called current mode architecture [1], where a (proportional-to-absolute temperature) PTAT current is summed up to a (complementary-to-absolute temperature) CTAT current to give a current that, at first order, is independent of the temperature. This current is then delivered to an output resistor (R3), which generates the output voltage $V_{ref,out}$. The internal operational amplifier has a single-stage folded cascode topology with an active load. A suitable start-up auxiliary circuit (not shown in this figure) provides the start of the bandgap when the power supply is switched on.

The output voltage is given by

$$V_{ref,out} = (R_3/R_2)*V_{be} + V_T*(R_3/R_1)*\ln(n) \tag{2.22}$$

From the above equation, $V_{ref,out}$ absolute accuracy is related to resistor matching and to the bipolar transistor's ratio n, so that in the real world, an error on the output absolute accuracy appears, usually in the order of some tenth of mV. To mitigate this unwanted effect, a trimming can be introduced, as shown in the figure. The calibration is achieved by acting on the resistors

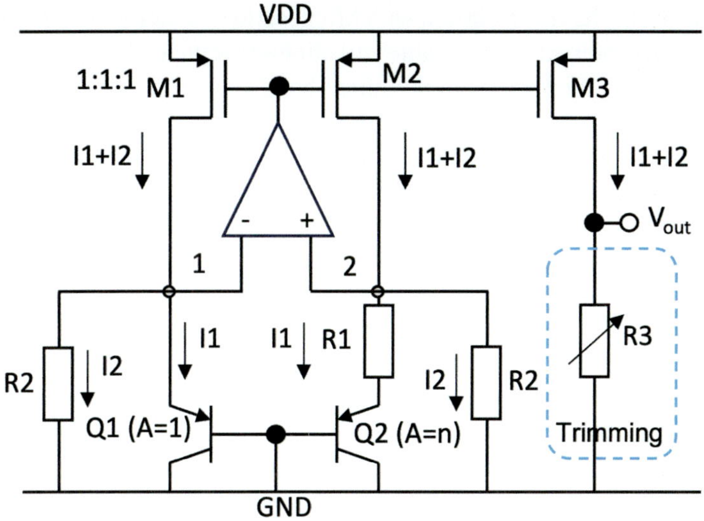

Figure 2.31 Simplified bandgap schematic.

of the structure, namely on R_3. The overall resistor is realized as a string of binary unit resistors, R_U, connected in series. The number of unit resistors is proportional to the precision required by the application. In our case, we decided on a resistor string with a fixed part and a trimmable part on 16 levels driven by 4 bits, as shown in Figure 2.32, according to the equation:

$$R_3 = R_{FIX} + 2^n * R_U \tag{2.23}$$

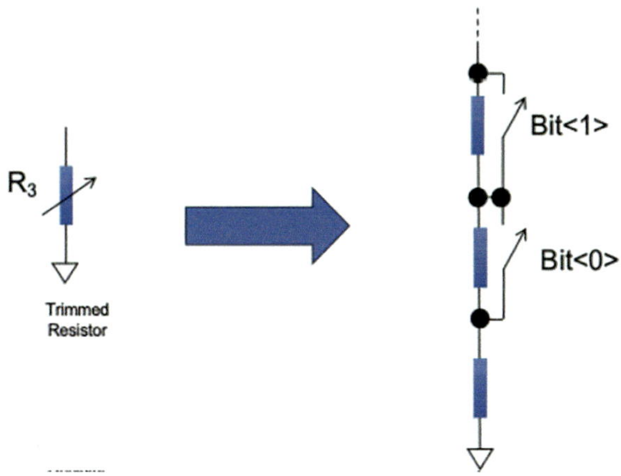

Figure 2.32 Detail of the trimming section of the BGRs.

The value of R_{FIX} has been chosen with Montecarlo simulations to be in between the inaccuracy interval. The variable part allows to regulate to the correct value.

In our test chip, we designed three bandgaps with fixed output (0.6 V, 0.9 V, and 1.2 V) and three with trimmable output (approximately 10 mV around target voltages of 0.6 V, 0.9 V, and 1.2 V). Trimming is activated by means of four external bits.

A suitable start-up circuit (Figure 2.33) has finally been added to avoid the BGR remaining in an off-by state when the power supply rises. Voltages V1 and V2 are connected to nodes 1 and 2 in Figure 2.31.

Finally, after each bandgap, an RC filter has been added (Figure 2.34), with benefits on forward noise, feedback noise, and analog single event transient (ASET) on BGR [47], [48]. The output of the filter can be used, for example, as input for an LDO voltage reference.

As concerns the layout of the bandgap, we applied all the best practices related to RHBD at the physical level. Namely, we designed ELT MOS (both nMOS and pMOS) against TID and extended guard-ring for inter-leakage and SEL. The resistors have been designed employing the silicide polysilicon as a material, which has been demonstrated to be insensitive to dose [17], [49]. To speed up the layout design and to improve the matching, we have adopted an approach like the one of the standard cells, by placing alternate rows of PMOS

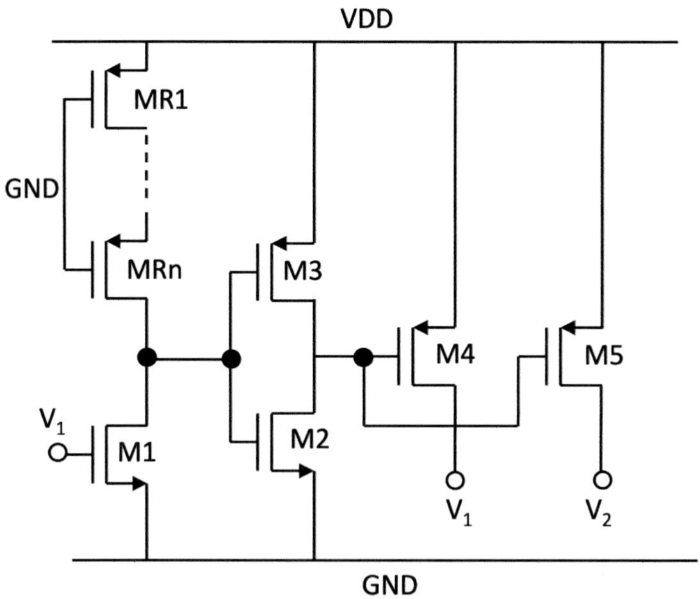

Figure 2.33 Start-up circuit of the BGR.

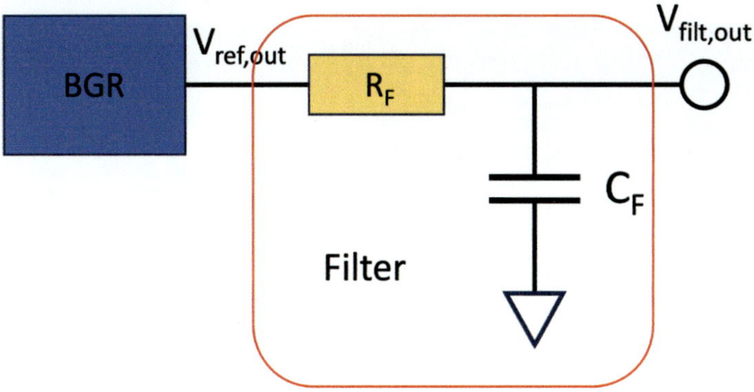

Figure 2.34 BGR with cascaded filter.

and NMOS devices, and closing the block at top, bottom, right, and left with substrate-well guard-rings. This strategy also has a further advantage, i.e., it helps to mitigate the other two unwanted effects. Indeed, guard-rings bring benefits both in terms of inter-device leakage (the generation of a conductive path induced by an accumulation of charges in the insulation between a MOS and the neighboring one) and of single events, in particular latch-up (single-event latch-up, SEL). For the latter effect, extended substrate bias and well bias rings have been inserted to separate PMOS and NMOS and provide isolation that makes latch-up due to collision with energetic ions unlikely.

The six BGRs have been integrated in a standard 3.3 V 180-nm CMOS process (XFAB XH018). The layout of two BGRs without and with trimming is shown in Figure 2.35(a) and (b).

The layout of the BGR opamp is shown in Figure 2.36. It is possible to identify the poly resistors, the MIM cap, and the rows of annular MOS. They are placed in a standard-cell fashion, as can be seen in Figure 2.37 (zoomed view of the input differential stage). The differential MOS has a common centroid placement to mitigate both offset (due to mismatch) and ASET [50]. Substrate or well guard-rings surround the MOS rows.

The design is modular and easily adjustable for different output voltages by acting on resistor R_3. The size of both untrimmed and trimmed BGR is 146×220 um^2, while with the insertion of the output filter is 146×247 mm^2, which are very compact areas.

Figure 2.38 shows the layout of the six BGRs included in the test chip.

The whole test chip assembled in a 32-pin ceramic dual-in-line package (CERDIP) is shown in Figure 2.39. It includes the six BGRs (in the center), together with other circuits (LDOs), which are not relevant for the present discussion.

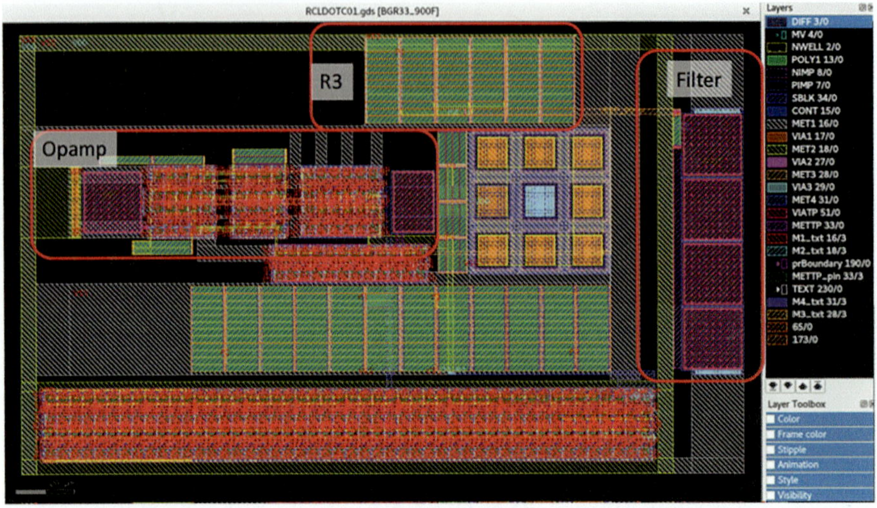

(a)

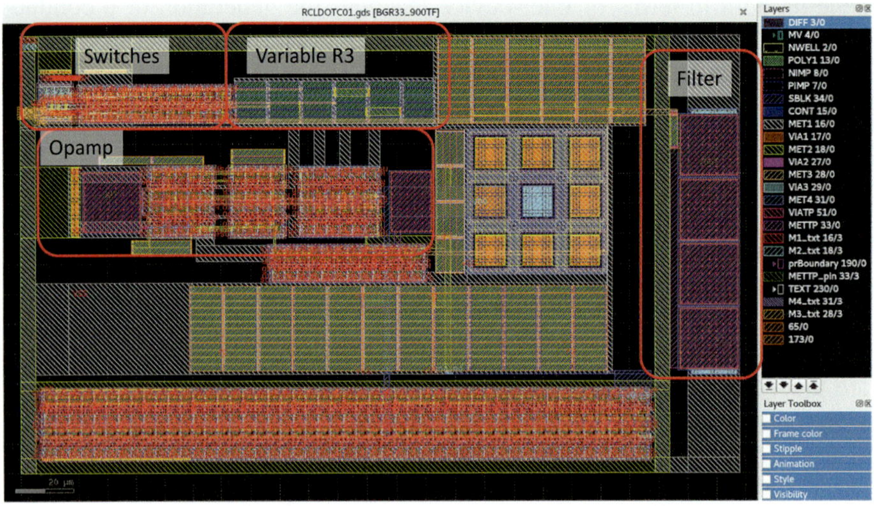

(b)

Figure 2.35 Layout of BGRs without trimming (a) and with trimming (b).

2.5.1 In-lab measurements

Each BGRs have been preliminarily characterized both in terms of static performance, namely absolute output voltage value, trimming capability, behavior with temperature (27 °C to 90 °C only), and power consumption.

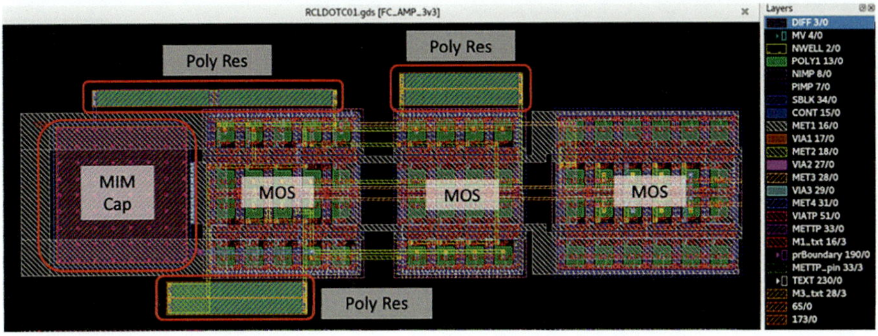

Figure 2.36 Layout of the opamp used in the BGR.

Figure 2.40 shows the measured output voltages for the three BGRs with fixed voltage on five samples. The spread due to process inaccuracy is 3% for 0.6 V, <4% for 0.9 V, and 8% for 1.2 V.

The regulation capability in the trimmable version is around 10 mV, as shown in Figure 2.41, which corresponds to 1% for 0.9 V BGR. From experimental measurements, it results too small to cover for process variability and will be increased in a future version. However, measurements confirm the validity of the principle.

The BGRs output voltage has also been measured versus the voltage supply. A ± 10% variation with respect to the nominal value of Vdd = 3.3 V

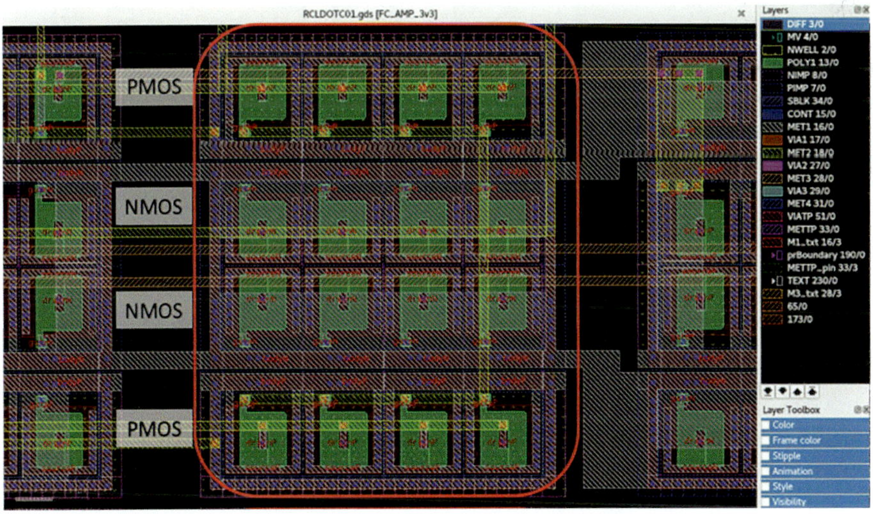

Figure 2.37 Zoomed view of the layout of the input differential pair in the BGR opamp.

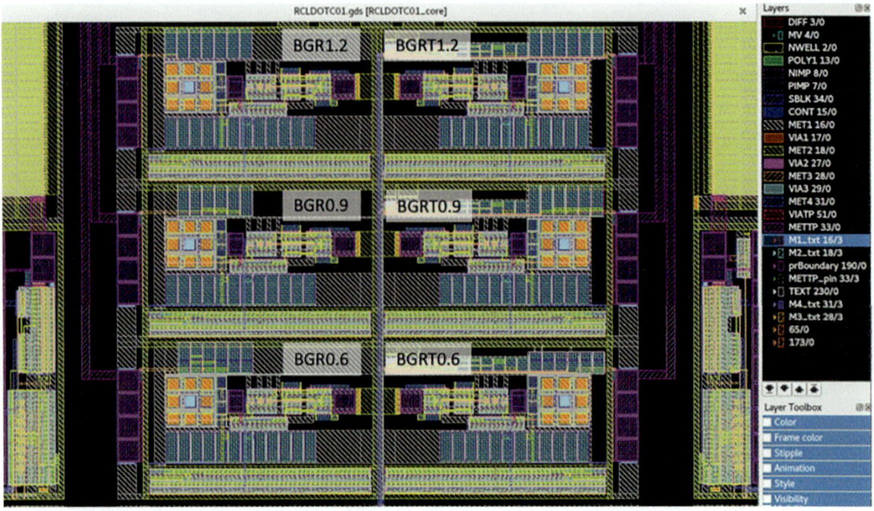

Figure 2.38 Layout of the six BGRs.

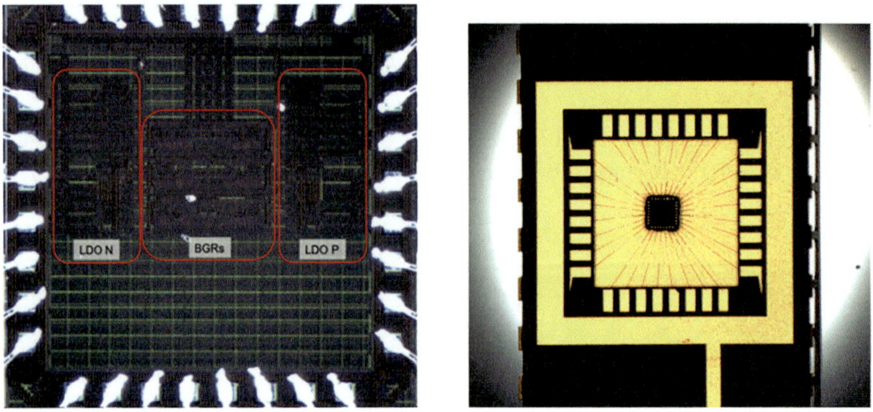

(a) (b)

Figure 2.39 Photo of the six BGRs test chip: (a) internal, (b) external.

has been shows these measurements assumed. Six samples of the test chip have been measured. Figure 2.42 shows measurements on sample 9.

The deviation of the output voltage with respect to the value at 3.3 V nominal is less than ± 2 mV, which means approximately 0.4%.

The BGRs output voltage has also been characterized versus temperature (Figure 2.43). The considered temperature range is only higher with respect to ambient temperature (30 °C–90 °C).

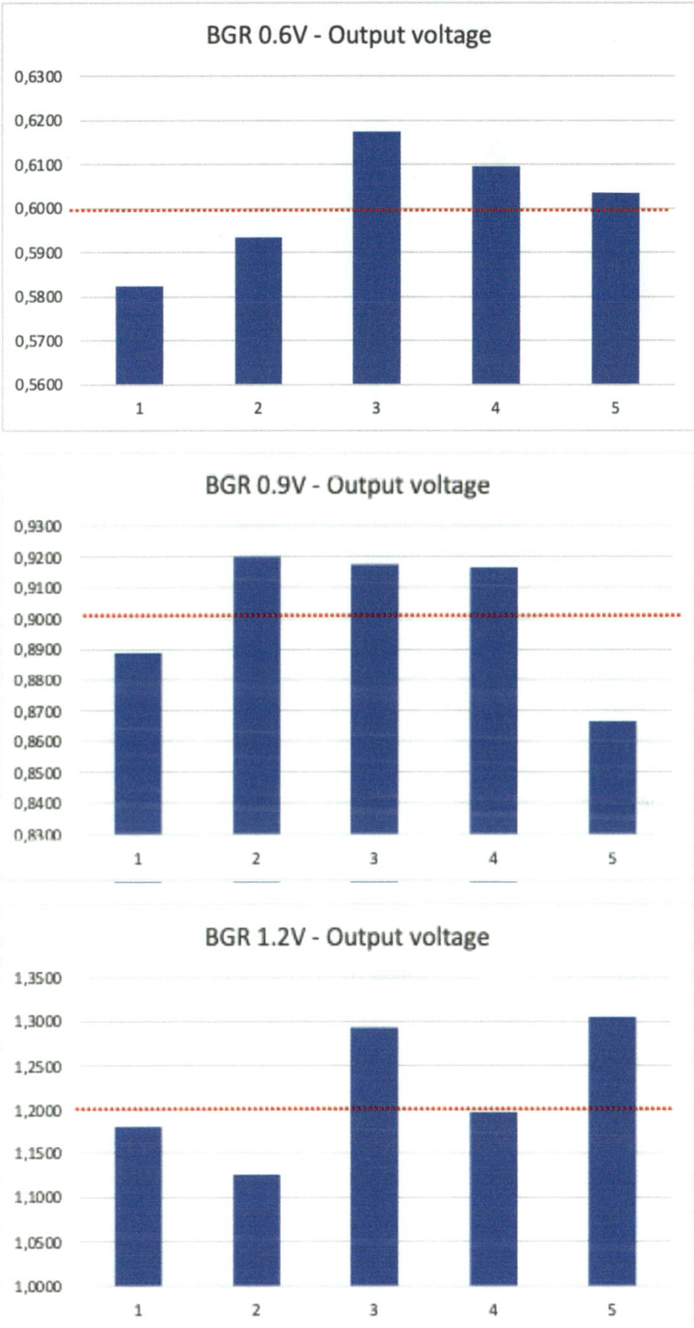

Figure 2.40 Measured BGRs fixed output voltage.

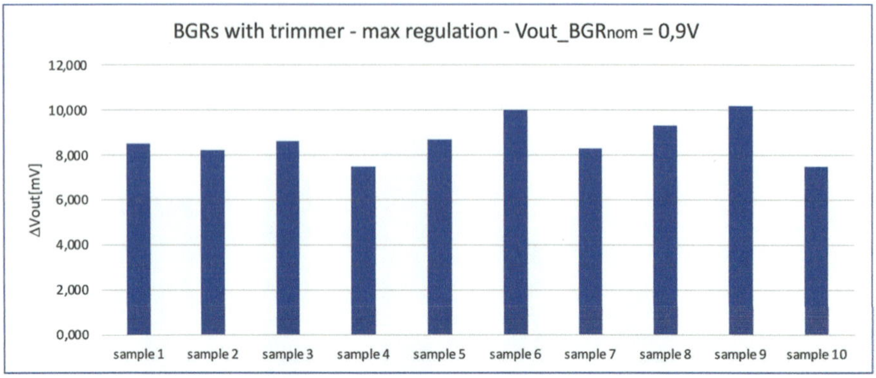

Figure 2.41 Regulation range of 0.9 V BGR.

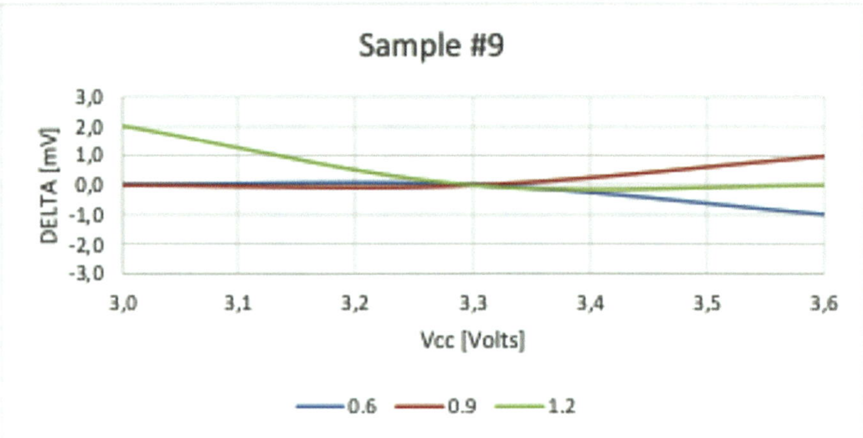

Figure 2.42 Variation of the BGRs output voltage with respect to the nominal value when Vdd ranges from 3 to 3.6 V.

The overall current consumption of the six BGRs is 690 uA, which can be approximately calculated as 115 uA each.

2.5.2 Measurements under radiation

The rad-hard BGRs have been tested under ^{60}Co at the University of Palermo facility using the RedCat Devices proprietary Leonida Platform, described in Chapter 5. Using a 0.75 rad/s dose rate, during the experiment, the BGRs reached up to 700 krad (Si) with no significant degradation in terms of electrical performance.

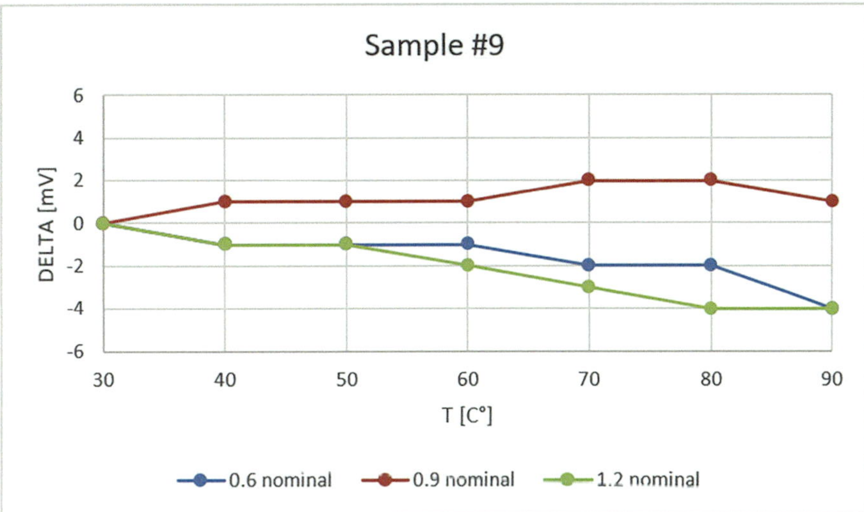

Figure 2.43 Variation of the BGRs output voltage with respect to the nominal value when the temperature ranges from 30 °**C** to 90 °**C**.

Due to the extreme environmental conditions, especially the high levels of radiation, it is crucial to develop an appropriate testing methodology. Traditional methods – such as using a test board with the device under test (DUT) mounted and lab equipment to generate signals and record outputs – are not viable. Only the DUT should be subjected to radiation, while measurement tools and auxiliary circuits must remain outside the irradiation zone. As a result, all measurement procedures must be conducted remotely.

As will be described in the dedicated Chapter 5, the proposed solution is based on splitting the test board into two boards: a mother-board (MB), which hosts all the controls, exposed, and a daughterboard (DB) with the DUT only. Here, as the mother-board, an off-the-shelf Arduino Due board plus a custom shield are used. The analog output of the proposed BGRs is converted into digital by a very accurate ADC onboard the Arduino Due, and then recorded. The custom shield contains the circuitry for the current sense and the needed power supplies. In the following, the measurement results obtained under TID with such methodology are presented.

For the TID measurements, we performed a series of 10 sequential measurements on the static current consumption and the output voltage of the BGRs by recording the results in a file. Figure 2.44 shows the 10 recorded measurements of the overall current consumption of the six BGRs from 0 rad

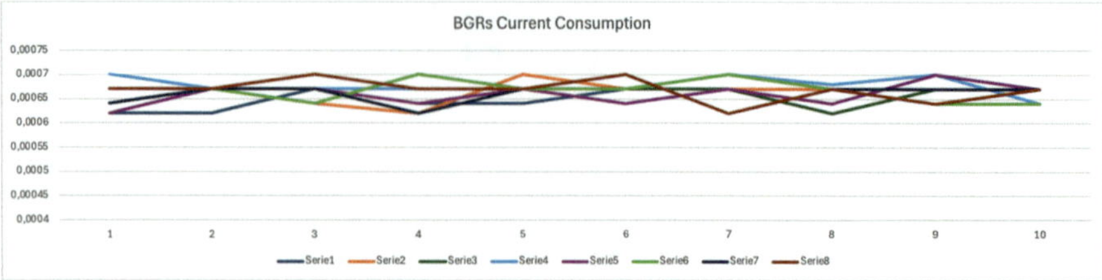

Figure 2.44 Ten sequential current consumption measurements of BGRs as a function of TID (Serie 1 corresponds to 0 rad, while Serie 8 to 700 krad).

(Si) to 700 krad (Si), with an interval of 100 krad (Si). The variation which can be observed between each measurement is due to the inaccuracy of measurements, intrinsic in the two boards approach, and in the noisy environment for TID measurements.

All measured values are between 620 uA and 700 uA and do not show any significant degradation (the nominal value was 690 uA).

The variation of the output voltage for 0.9 V fixed BGR on 10 sequential runs as a function of total dose is shown in Figure 2.45. As before, the

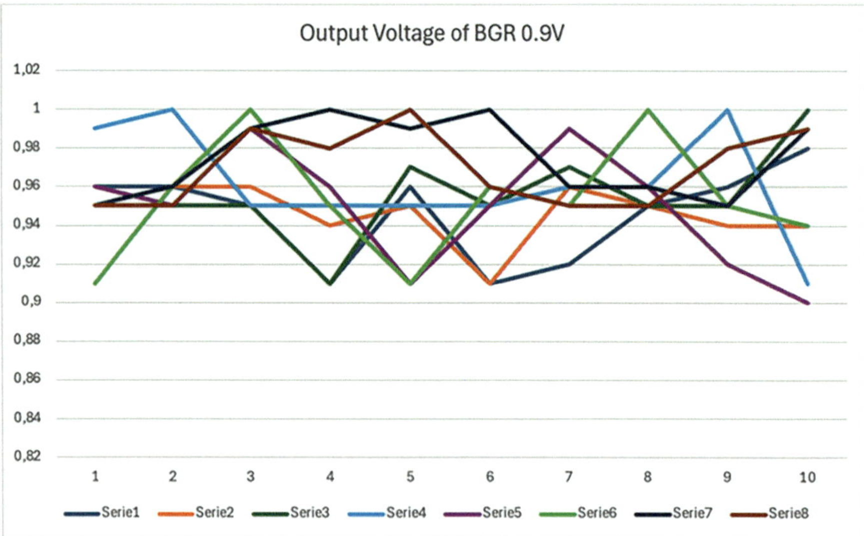

Figure 2.45 Ten sequential output voltage measurements of 0.9 V fixed BGR as function of TID (Serie 1 corresponds to 0 rad, while Serie 8 to 700 krad).

variation which can be observed between each measurement (± 40 mV) is due to measurement inaccuracy, intrinsic in the two boards approach, and in the noisy environment for TID measurements.

The trend of the output voltage does not show a significant degradation (nominal voltage between 0.91 V and 0.95 V). The same behavior has been observed for 0.6 V and 1.2 V fixed BGRs.

Finally, we performed the heavy ion test at RADEF University of Jyvaskyla (Finland) on a couple of samples of the test chip. The characteristics of the beam were:

- Heavy Ion: Xenon
- Energy: 2059 MeV
- Flux: 9000 #ions/s
- LET: 60 MeV*cm^2/mg (Si)

The environment is less noisy, and the sequential measurements (up to 50) have been more stable. Figure 2.46 shows the 50 sequential measurements of the 0.9 V fixed output voltage BGR. Any SET appeared.

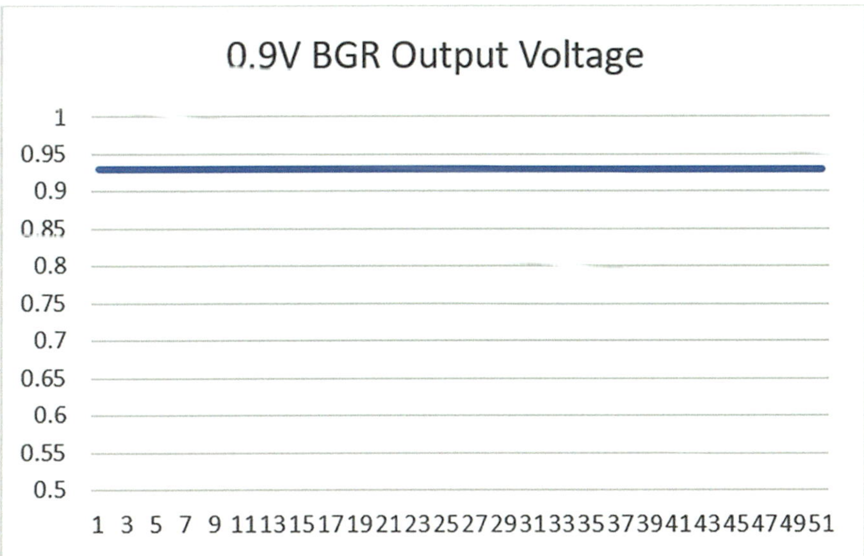

Figure 2.46 Fifty sequential output voltage measurements of 0.9 V fixed BGR during heavy ions test.

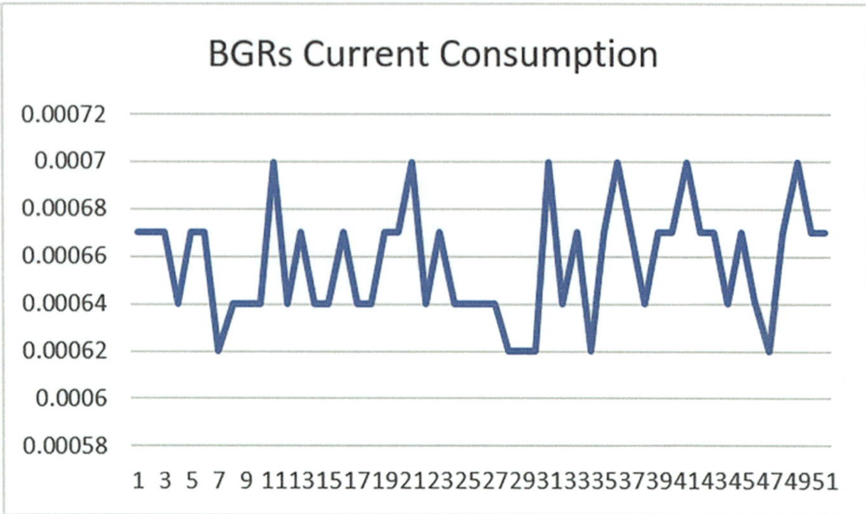

Figure 2.47 Fifty sequential overall current consumption measurements of BGRs during heavy ions test.

Figure 2.47 shows the 50 sequential measurements of the overall current consumption of BGRs. Any SEL (sudden and uncontrolled increase of the current) appeared.

2.6 Conclusion

In this chapter, we have analysed the basic concept of voltage references, which is an auxiliary block commonly used in several complex systems. It is important to study how to design these blocks appropriately since their performance affects speed, dynamic range, power consumption, etc. of the circuits where they are inserted. This is particularly true in harsh environment electronics, where, besides the usual causes of degradation (temperature, power supply variations, marching, etc.), also resilience to radiation plays a crucial role. Thus, we illustrated an overview of the main design techniques proposed in the literature to overcome this issue, to give the reader useful design insights and an almost complete reference list with the most recent proposals.

Finally, we described the design and experimental results in the lab, under TID and under heavy ions, of six rad-hard bandgap voltage references (BGRs) proposed by the authors. They have been implemented in a conventional 180-nm CMOS process using RHBD techniques. Measurements do not show any significant degradation up to 700 krad (Si) dose, and under heavy ions, any latch-up phenomenon has been observed.

References

[1] F. Maloberti, "Analog Design for CMOS VLSI Systems", Springer, 2001.
[2] C. W. Kok, W. S. Tam, "CMOS Voltage References", Wiley-IEEE Press, 2013.
[3] G. A. Rincon-Mora, "Voltage References", Wiley-IEEE Press, 2000.
[4] D. M. Colombo et al., "Sub-1 V band-gap based and MOS threshold-voltage based voltage references in 0.13 lm CMOS", Analog Integr. Circ. and Sig. Process., Jan. 2014, pp. 25-37.
[5] R. Baker, "CMOS Circuit Design, Layout and Simulations- 3rd Ed.", Wiley, 2010.
[6] H. Banba, et al., "A CMOS bandgap reference circuit with sub-1-V operation", Jour. Of Solid-State Circuits, Vol. 34, pp. 670-674, May 1999.
[7] S. Liu, J. Baker, "Process and Temperature Performance of a CMOS Beta-Multiplier Voltage Reference", 1998 Midwest Symposium on Circuits and Systems, MWSCAS, Notre Dame, USA, 1998.
[8] X. Jing, et al., "A 0.5V nanoWatt CMOS Voltage Reference with Two High PSRR Outputs", 2012 IEEE International Symposium on Circuits and Systems (ISCAS), Seoul, 2012, 2837-2840.
[9] K. Ueno, et al., "A 300 nW, 15 ppm/ C, 20 ppm/V CMOS Voltage Reference Circuit Consisting of Subthreshold MOSFETs", IEEE Journ. of Solid-State Circuits, Vol. 44, No. 7, July 2009, pp. 2047-2054.
[10] K. Kruckmeyer, et al., "Impact of Reference Voltage on the ELDRS Characteristics of the LM4050 Shunt Voltage Reference", RADECS 2011, pp. 926-930.
[11] S. S. McClure, et al., "Total Dose Performance of Radiation Hardened Voltage Regulators and References", IEEE Radiation Effects Data Workshop. NSREC 2001.
[12] J. S. George, J. R. Srour, et al., "SEE and TID Effects in Transistors and Voltage", 2016 IEEE Radiation Effects Data Workshop (REDW).
[13] G. Cressler and H. Mantooth Ed., "Extreme Environment Electronics", CRC Press, 2013.
[14] R. L. Pease, "Total ionizing dose effects in bipolar devices and circuits,", IEEE Trans. Nucl. Sci., vol. 50, no. 3, pp. 539–551, Jun. 2003.
[15] Y. Cao et al., "A 4.5 MGy TID-Tolerant CMOS Bandgap Reference Circuit Using a Dynamic Base Leakage Compensation Technique", IEEE Trans. on Nucl. Science, Vol. 60, No. 4, pp. 2819-2824, August 2013.

[16] V. Gromov et al., “A Radiation Hard Bandgap Reference Circuit in a Standard 0.13um CMOS Technology”, IEEE Trans. on Nucl. Science, Vol. 54, No. 6, Dec. 2007.

[17] R. Pease, “The design of band-gap reference circuits: Trials and tribulations”, in Proc. IEEE 1990 Bipolar Circuits Technology Meeting, 1990, pp. 214–218.

[18] K. Shetler et al., “Radiation Hardening of Voltage References UsingChopper Stabilization”, IEEE Trans. on Nucl. Science, Vol. 62, No. 6, Dec. 2015.

[19] T. Vergine et al., “A 65 nm Rad-hard Bandgap Voltage Reference for LHC Environment”, IEEE Trans. on Nucl. Science, Vol. 63, No. 3, June 2016.

[20] R. C. Lacoe, “Improving integrated circuit performance through the application of hardness-by-design methodology,” IEEE Trans. Nucl. Sci., vol. 55, no. 4, pp. 1903–1925, Aug. 2008.

[21] M. Kuczynska, et al., “Development of Radiation-hard Bandgap Reference and Temperature Sensor in CMOS 130 nm Technology”, Proc. 22nd Int. Conf. “Mixed Design of Integrated Circuits and Systems”, June 25-27, 2015, Poland.

[22] B. M. McCue, et al., “A Wide Temperature, Radiation Tolerant, CMOS-Compatible Precision Voltage Reference for Extreme Radiation Environment Instrumentation Systems”, IEEE Trans. on Nucl. Science, Vol. 60, No. 3, June 2013.

[23] A. J. Annema, “Low-power bandgap references featuring DTMOST’s,” IEEE J. Solid-State Circuits, vol. 34, no. 7, pp. 949–955, Jul. 1999.

[24] Y. Piccin et al., “Radiation-Hardening Technique for Voltage Reference Circuit in a Standard 130nm CMOS Technology” IEEE Trans. on Nucl. Science, Vol. 61, No. 2, April 2014.

[25] G.M. Groeseneken, J. P. Colinge, H. E.Maes, J. C. Alderman, and S. Holt, “Temperature dependence of threshold voltage in thin-film SOI MOSFETs,” IEEE Electron Device Lett., vol. 11, no. 8, pp. 329–331, Aug. 1990.

[26] T. R. Oldham and F. B. McLean, “Total ionizing dose effects in MOS oxides and devices,” IEEE Trans. Nucl. Sci., vol. 50, no. 3, pp. 483–499, Jun. 2003.

[27] C. Andreou, C. Calligaro, J. Georgiou et al., “A Subthreshold, Low-Power, RHBD Reference Circuit, for Earth Observation and Communication Satellites “, ISCAS 2015, 2015.

[28] D. M. Colombo, et al., "Total Dose Effects on Voltage References in 130-nm CMOS Technology", IEEE Trans. on Device and Materials Reliability, Vol. 18, No. 1, March 2018, pp. 27-36.

[29] G. Traversi, et al., "Characterization of bandgap reference circuits designed for high energy physics applications", Nuclear Instruments and Methods in Physics Research A 824, 2016, pp. 371–373.

[30] D. Fusco, T. R. Balen, "Radiation Effects in Low Power and Ultra Low Power Voltage References", 17th IEEE Latin-American Test Symposium - LATS 2016, pp. 111-116.

[31] A. Hassan et al., "Ultra-low Power CMOS Voltage Referencefor High Temperature Applications up to 300°C", Electronics, Circuits, and Systems (ICECS), 2015 IEEE International Conference on, 2015.

[32] A. Fitas, N. Horta, J. Guilherme, "Design of a Radiation-Hardened Curvature Compensated Bandgap Reference Circuit", 2016 12th Conference on Ph.D. Research in Microelectronics and Electronics (PRIME), Lisbon, 2016.

[33] U. Gatti, C. Calligaro, E. Pikhay, Y. Roizin "Radiation-hardening methodologies for Flash ADC", Analog Integrated Circuits and Signal Processing Journal (Springer), DOI 10.1007/s10470-015-0627-7.

[34] Q. Zhao et al., "A single-event transient mitigation technique for bandgap reference utilizing in space application", Authorea Preprints, 2023.

[35] Y. Ren, L. Chen, and J. Bi, "An rhbd bandgap reference utilizing single event transient isolation technique", IEEE Trans. Nucl. Sci., vol. 63, no. 3, pp. 1927–1933, 2016.

[36] J. Liu, D. Wang, B. Liang, Y. Chi, D. Luo, and S. Xu, "Soft error tolerant bandgap reference utilizing single-event transient filtering technique", IEEE Trans. Nucl. Sci., vol. 71, no. 4, pp. 895–901, 2024.

[37] C. Peng, et al., "A Low Overhead Single Event Transient Tolerant Bandgap Reference Based on Radiation-Hardened-by-Design (RHBD) Technique", IEEE Transactions on Aerospace and Electronic Systems (Early Acess), 19 June 2025, pp. 1-9.

[38] C. Andreou, C. Calligaro, J. Georgiou et al., "Single Event Transients and Pulse Quenching Effects in Bandgap Reference Topologies For Space Applications", IEEE Transactions on Nuclear Science Volume: 63, Issue: 6, December 2016.

[39] P. Carniti, et al., "A Multifunction Radiation-Hardened HV and LV Linear Regulator for SiPM-Based HEP Detectors", Published in: IEEE Tranc. on Nucl. Science, Vol. 71, No. 7, pp. 1399-1408, July 2024.

[40] E. Boufouss et al., "High Temperature and Radiation Hard CMOS SOI Sub-threshold Voltage Reference", SOI-3D-Subthreshold Microelectronics Technology Unified Conference (S3S), 2013.
[41] J-P. Colinge, Silicon-on-Insulator technology: materials to VLSI. 2nd ed., Boston: Kluwer Academic Publishers, 1997; pp. 223-237.
[42] L. Najafizadeh, et al., "Proton Tolerance of SiGe Precision Voltage References for Extreme Temperature Range Electronics", IEEE Trans. on Nucl. Science, Vol. 53, No. 6, pp. 3210-3216, Dec. 2006.
[43] L. Najafizadeh et al., "A Comparison of the Effects of X-Ray and Proton Irradiation on the Performance of SiGe Precision Voltage References", IEEE Trans. on Nucl. Science, Vol. 54, No. 6, pp. 2238-2244, Dec. 2007.
[44] J. D. Cressler, "Radiation effects in SiGe technology," IEEE Trans. Nucl. Sci., vol. 60, no. 3, pp. 1992–2014, Jun. 2013.
[45] A. S. Cardoso, et al., "Single-Event Transient and Total Dose Response of Precision Voltage Reference Circuits Designed in a 90-nm SiGe BiCMOS Technology".
[46] A. Rahman et al., "High-Temperature Voltage and Current References in Silicon Carbide CMOS", IEEE Trans. on Electronic Devices, Vol. 63, No. 6, June 2016.
[47] J. Chen, et al., "ASET and TID Characterization of a Radiation Hardened Bandgap Voltage Reference in a 28-nm Bulk CMOS Technology", IEEE Trans. on Nucl. Science, Vol. 69, No. 5, pp. 1141-1147, May 2022.
[48] N. Lourenco, et al., "Radiation-Hardened Bandgap Voltage and Current Reference for Space Applications with 2.38ppm/°C Temperature Coefficient", 2022 18th International Conference on Synthesis, Modeling, Analysis and Simulation Methods and Applications to Circuit Design (SMACD).
[49] C. L. Axness, et al., "Radiation Characteristica of SIPOS and Polysilicon Resistors", IEEE Trans. on Nucl. Science, Vol. 38, No. 6, Dec. 1991, pp. 1365-1369.
[50] A. T. Kelly, et al., "Differential Analog Layout for Improved ASET Tolerance", IEEE Trans. on Nucl. Science, Vol. 54, No. 6, pp. 2053-2059, Dec. 2007.

3

Rad-hard Digital-to-Analog Converter (DAC)

Abstract

Analog-to-digital (ADC) and digital-to-analog (DAC) converters are fundamental components for bridging the analog to the digital world. Data converters perform a transformation of signals: from continuous-time and continuous-amplitude to discrete-time (at a sampling frequency f_s) and quantized-amplitude (and vice versa). They find wide application in systems where huge digital processing units have to exchange data with the external world, which is analog in nature. Space applications are not an exception, thus making data converters key components in virtually all on-board electronic circuits and systems. The characteristics of the converters, especially in terms of resolution and speed, have a tremendous impact on the performance of the systems, which depend on how much accurate the conversion (in both directions) of the analog signals is. Popular signal examples are those coming from temperature sensors or those fed to propulsion controls. In such cases, having ADC/DAC that provide very precise conversion under any extreme condition (such as under radiation exposure) is mandatory to avoid mission failures.

In this chapter, focused on DAC, we first recall its specifications, which give general information about the features and limits of the static and dynamic operations. Then, we will illustrate the basic principles of DAC and the three main topologies. Starting from the latter, we will show which are the most promising rad-hard DACs that have been presented in literature, starting from known solutions and applying rad-hard design provisions. We also give a look at more innovative solutions for specific applications. Finally, two rad-hard digital-to-analog converters recently implemented by the Authors are described in detail.

DOI: 10.1201/9788743813286-3

3.1 Introduction

Due to the increased need for radiation-tolerant analog and mixed-signal circuits in recent years, the development of precision analog circuits has become more important. For instance, as an interface between digital and analog worlds, the digital-to-analog converter (DAC), which converts the signal from digital logical expressions "0" and "1" to a continuous amplitude value of the analog domain, is an essential module in the whole electronic system. Currently, DACs and ADCs (see Chapter 4) are very prevalent in modern electronic devices. However, their performance is often a bottleneck for practical applications, because of the non-idealities they introduce, such as poor accuracy, speed limitation, distortions, etc. For terrestrial applications, many high-speed DAC specifications are generated by the video market needs (i.e., video signals from a digital source, such as a computer or a DVD, or for TV). DACs are also used for wired and wireless communications that use a variety of modulation and encoding schemes to exchange digital information. Headphones play back the sound after the DAC conversion. Since in this case the signal bandwidth is relatively low, DACs used for audio applications normally exploit oversampling. In space, the DAC converter represents a fundamental component as a stand-alone device and as an embedded macro in a complex system (such as microcontrollers used in satellite as housekeeping or for propulsion control, etc.).

The goal for "space" engineers is to design and manufacture high-performance radiation-tolerant or even radiation-hard DAC and ADC by using standard commercial processes of integration that are inexpensive, with easy maintenance, and widely available in the industry.

3.2 Digital-to-Analog Converters: Basic Principles

The digital-to-analog conversion is a fundamental function in data processing systems, where a digital processing system must interface with the analog world (i.e., sensors, actuators, etc.). DACs are also used inside multi-step analog-to-digital converters for the residual reconstruction. In practice, an ideal DAC receives N bits as input and produces either voltage or current as an analog output signal capable of driving a load (Figure 3.1). In the latter case, the output current is often transformed into a voltage using termination resistors.

The analog output of the DAC is often a continuous-time sample and a held representation of the wanted signal. It is assumed that a separate external function (low-pass filter and Sinc correction) realizes the reconstruction,

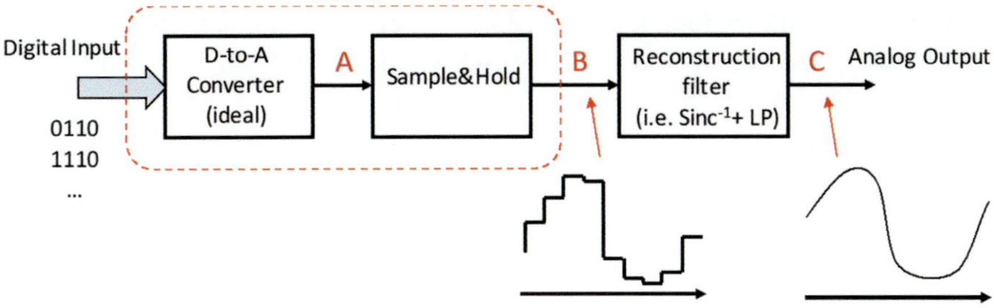

Figure 3.1 DAC principle and related waveforms.

i.e., filters out the high-frequency components (see spectra in Figure 3.2). So, the overall output bandwidth lies within the first Nyquist zone.

The input binary code of an N-bit DAC starts from the bit b_0 called LSB (least significant bit), while it ends with the bit b_N, which is called MSB (most significant bit). The so-called Nyquist DACs generate a series of outputs that have a one-to-one correspondence with the input code and operate at a sample frequency that is twice the maximum input bit rate. The analogue output signals are derived from the digital inputs via one (or two) reference voltage (or current) V_{ref}:

$$V_{out} = \left(V_{ref} / 2^N\right)\left(2^0 b_0 + 2^1 b_1 + 2^2 b_2 + 2^3 b_3 + ::: + 2^{N-1} b_{N-1}\right) \tag{3.1}$$

V_{ref} defines the dynamic range of the converter and is often known as V_{FS} (full-scale voltage).

It is useful to define the voltage of the LSB (V_{LSB}), which is the voltage corresponding to the smallest variation in the inputs that the DAC can convert. The ideal V_{LSB} is given by:

$$V_{LSB} = V_{FS} / 2^N \tag{3.2}$$

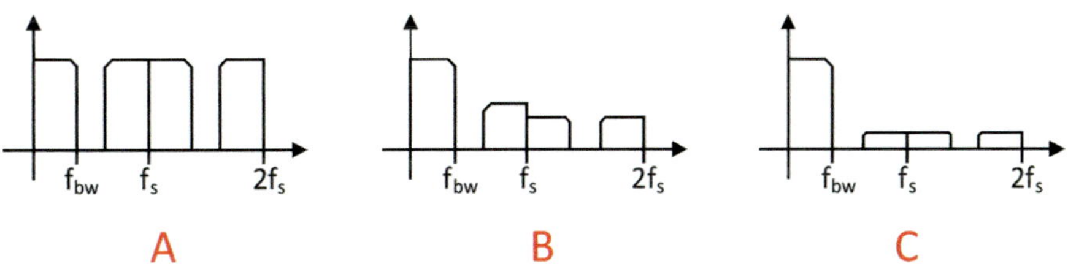

Figure 3.2 DAC output signal and reconstructed analog signal after a low-pass filtering.

With this voltage value, it is possible to define a unit called LSB. The maximum value that a DAC can assume when all bit b_i are "1" is $V_{out,max} = V_{FS}(2^N - 1)/2^N = V_{FS} - V_{LSB}$.

A detailed description of the properties of each block that builds up a DAC is not included in this book, since it is not the focus of this publication, and it has already been treated widely in previous literature. The Readers can refer to Refs. [1]-[4] for deeper information.

Many characteristics define a DAC's performance. In the next section, each specification will be discussed before we look at the basic architectures.

3.2.1 DAC main performance

DAC performance is described by means of some specifications. Main parameters are [1]:

Static

- Resolution
- Offset
- Gain error
- Accuracy
- INL/DNL
- Monotonicity

Dynamic

- Settling time
- Sampling rate
- SNR
- ENOB/SINAD
- SFDR

The first static specification considered is the resolution of the DAC. Resolution is defined as the number of separated analogue levels corresponding to the different digital codes. In an N-bit DAC, the resolution will be 2^N levels. The ideal transfer curve is represented in Figure 3.3.

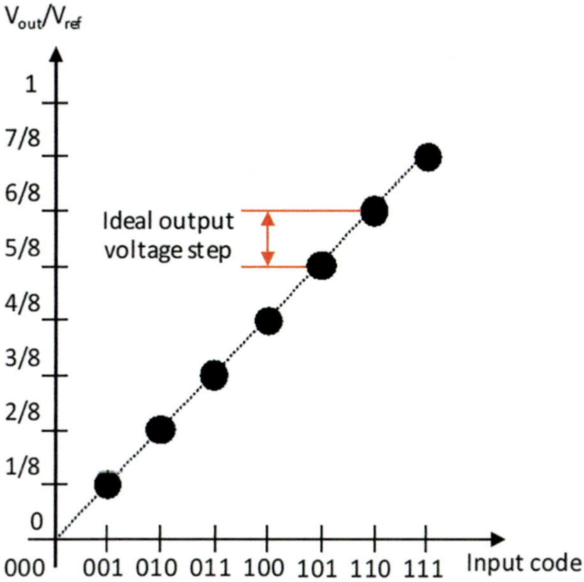

Figure 3.3 Ideal transfer curve of a DAC.

The second characteristic is the offset, which is defined as the deviation of the actual analog output voltage from the ideal 0V when a zero-code input is applied (in case of a DAC with an output ranging from 0 to $V_{FS}-V_{LSB}$):

$$V_{off}(\text{LSB}) = \left(\frac{V_{out}}{V_{LSB}}\right)_{0::000} \tag{3.3}$$

The offset is expressed in units of LSB (Figure 3.4(a)). The third parameter is the gain error (V_{gain}), defined as the difference between the ideal full-scale value, V_{FS}, and the real full-scale value when the offset has been removed (Figure 3.4(b)):

$$V_{gain}(\text{LSB}) = \left(\frac{V_{out}}{V_{LSB}}\right)_{1::111} - \left(\frac{V_{out}}{V_{LSB}}\right)_{0::000} - \left(2^N - 1\right) \tag{3.4}$$

The fourth specification is the accuracy. For this, we must consider absolute accuracy and relative accuracy. The absolute accuracy is defined as the difference between the expected and the real transfer responses (analog levels that occur for each of the input digital codes), including the gain error, the offset

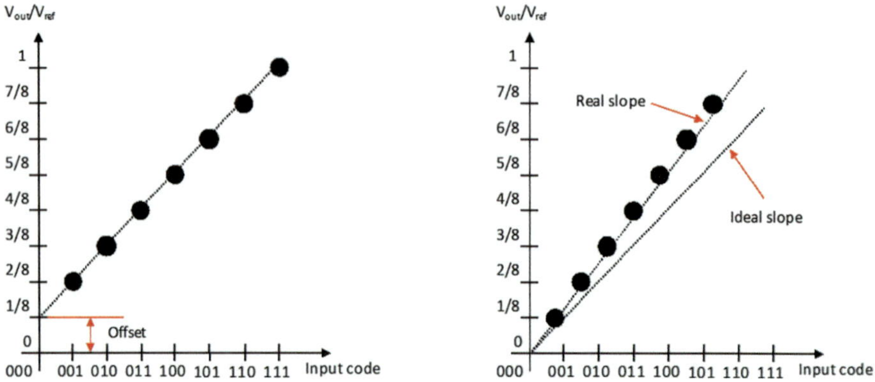

(a) (b)

Figure 3.4 DAC offset error (a) and gain error (b).

error, and the linearity error. The relative accuracy is defined as the absolute accuracy, with the exception that the errors of gain, offset, and linearity are not considered. The accuracy can be expressed in percentage of V_{FS}, effective number of bits (ENOB), or as a fraction of LSB.

The fifth parameters are the integral non-linearity (INL) and the differential non-linearity (DNL), shown in Figure 3.5(a) and (b), respectively. The former is defined as the deviation of the DAC response from a straight line (the ideal interpolating line). There are two ways to define a straight line. The first way is to use the endpoints of the transfer response, while the second way is to find the best-fit straight line such that the maximum difference is minimized. Instead, the INL is defined in units of LSB. The DNL is defined as the deviation of the real analog steps from the ideal width of the bin.

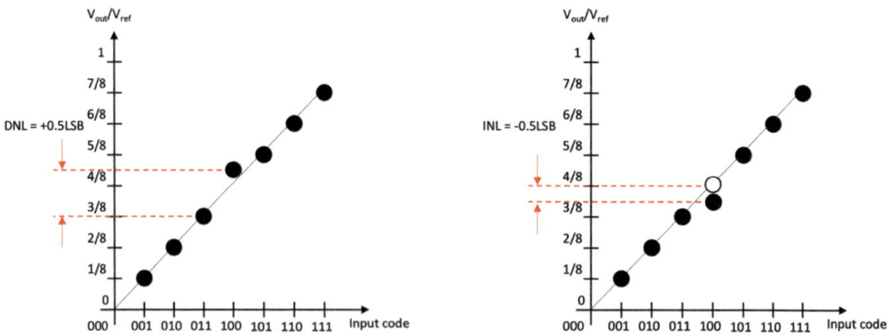

(a) (b)

Figure 3.5 DAC DNL (a) and INL (b) definitions.

The sixth specification is the monotonicity. A DAC is monotonic if its output signal increases with the input bit code increase. In other words, a DAC is monotonic if the slope of the transfer response has a definite sign, that is, whether increasing or decreasing. This ends the main static DAC characteristics.

The first dynamic parameter is the settling time, which is defined as the time the converter output takes to settle within some specified amount of the final value, depending on resolution. Instead, the sampling rate is defined as the rate at which the samples can be converted continuously. This quantity is almost equal to the inverse of the settling time.

The second dynamic parameter is the signal-to-noise ratio (SNR), which is defined as the ratio of the root-mean-square of the signal power (generally a sinewave) to the root-mean-square of the noise power at the analog output. The approximation or "rounding" effect in a converter is called "quantization," and the difference between the digital input and the analog output is referred to as "quantization error." In an ideal DAC converter where only the quantization noise is present and under some assumptions, for a full-scale sinewave signal, the SNR (over the entire Nyquist interval) is given by [2]:

$$\mathrm{SNR} = 6.02 * N + 1.76\ (\mathrm{dB}) \tag{3.5}$$

where N is the bit number. In a real converter, the noise power shall also include the noise (thermal, jitter, etc.) of the circuits.

Another important dynamic parameter is the SINAD (signal-and-distortion ratio). It is the ratio between the root-mean-square of the signal and the root-mean-square of the harmonic plus noise. The related effective-number-of-bit (ENOB) is a measure of the signal-to-noise and distortion ratio using bits. SINAD in dB and ENOB are related by:

$$\mathrm{ENOB} = \frac{\mathrm{SINAD}_{\mathrm{dB}} - 1.76}{6.02} \tag{3.6}$$

Finally, spurious-free dynamic range (SFDR) is the ratio of the root-mean-square output signal amplitude to the root-mean-square value of the highest spurious spectral component in the first Nyquist band.

3.2.2 Main DAC architectures

Based on the mean used to convert the digital signal to analog, several DAC architectures, each addressing different resolution-speed requirements, have been proposed. Precise DACs are built up starting from accurate elements available in a standard CMOS technology, i.e., resistors, capacitors, and MOS transistors. In the following, common overview material and basic principles behind their operation will be reviewed, focusing on the most popular topologies and on their pros and cons. A complete description of all variants of these circuits is outside the scope of this book and can be easily found in existing literature [2]-[4]. The most popular DAC topologies are:

- Resistive DAC
- Capacitive DAC
- Current-steering DAC

Resistive DAC

The approaches used to design a resistive DAC are to create, in some way, 2^N reference voltages employing resistors as basic elements. According to the digital input code, one of these voltage taps is selected as the analogue output signal.

In principle, this type of DAC is implemented with a string of 2^N resistors biased by V_{ref}, which divides the reference voltage into 2^N voltage values (Figure 3.6). A network of switches, driven by the digital input code, allows the selection of the corresponding voltage value and feeds it to the output or to the input terminal of a decoupling buffer. In the latter case, the buffer output terminal will be the DAC output terminal. In a CMOS technology, generally a complementary switch is used, that is, a switch formed by an nMOS and a pMOS. This represents the simplest way to realize a resistor string DAC.

The monotonicity of this DAC is guaranteed because any tap of the resistor string has a lower voltage than its upper neighbor tap. The accuracy depends on the matching of the resistors and the material they are made of. However, with the increase in the number of bits, the number of resistors, switches, and control lines becomes too large. Moreover, the parasitic load at the buffer input increases and is very variable with code. A further limitation of this circuit is speed, as the resistor string and the switches introduce a time constant and thus delays in the conversion process.

In order to reduce the capacitance load, it is possible to use a matrix topology called a folded resistor string (or X-Y addressing scheme). An example of it is shown in Figure 3.7. This solution is also used to reduce the decoder size and the number of control lines.

The 16 resistors (for 4-bit) are divided into four lines and arranged with a serpentine shape connected between V_{ref+} and V_{ref-} (Figure 3.7). Each line

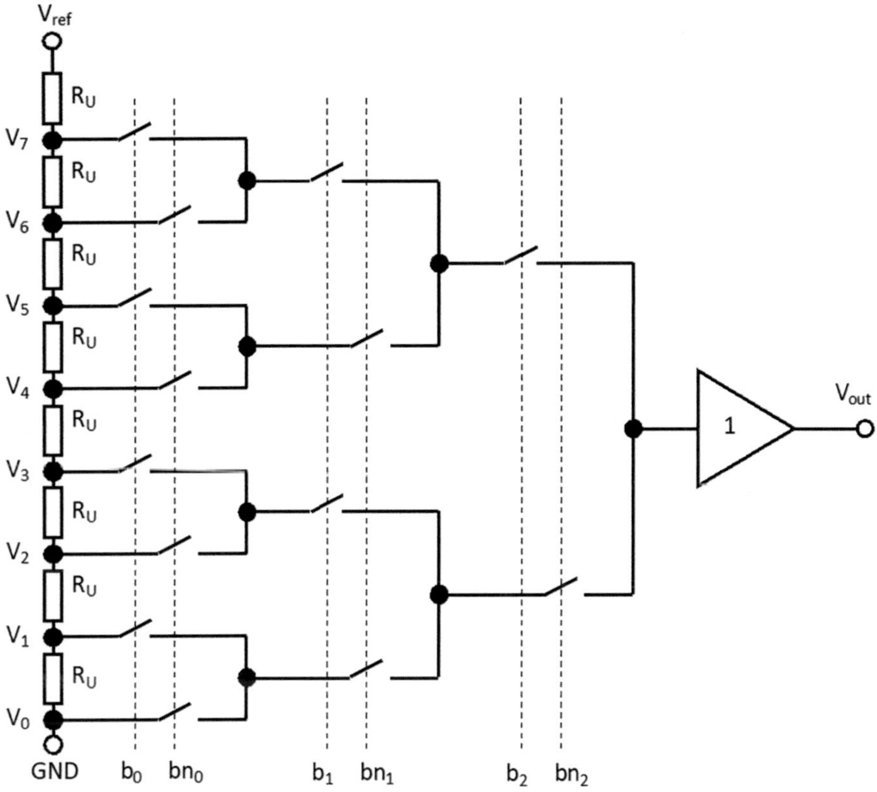

Figure 3.6 Resistor string DAC (Kelvin divider for 3-bit).

is selected by one of the decoded MSBs of the input code. The Y_i are the output of a decoder that converts the LSBs of the input code, and the X_i are the output of a decoder that converts the MSBs of the input code. The most significant bits select a column of the matrix. That is, a block of adjacent resistors is connected to switches controlled by less significant bits. One of these resistors is connected to the buffer by closing a further switch. The voltage at the terminals of each resistance is a fraction of the reference voltage ($V_{ref+} - V_{ref-}$), so the output voltage of the DAC is one of these voltages. Since there are 2^N resistances, 2^N voltage values are produced. You need to have a switch for each voltage value obtained, and since there are 2^N voltage values, you have that the number of transistors is: $2^*(2^N)/2$. The factor two derives from the fact that each switch, being a complementary switch, is composed of two transistors. The factor 1/2 instead results from the fact that only half of the transistors are connected to the output. Therefore, the number of transistors connected to the buffer is 2^N [1].

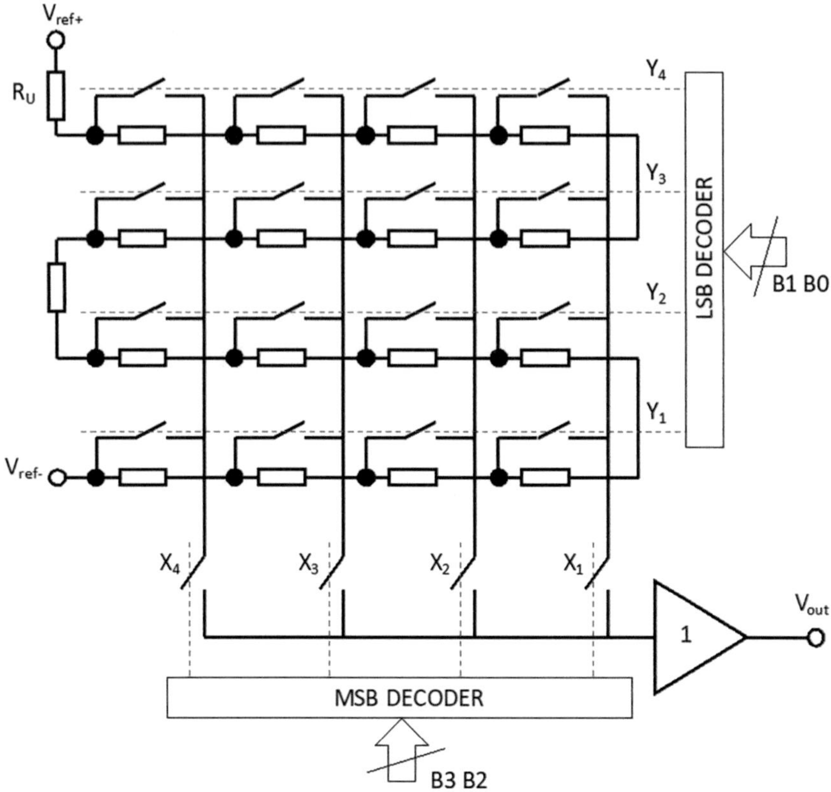

Figure 3.7 Matrix resistor string (folded resistor string for 4-bit).

A limit of the resistive divider DAC (in all its variants) is that the number of resistors increases exponentially with the number of bits. This problem can be solved with a so-called R-2R or reduced-resistors-ratio ladder (Figure 3.8,

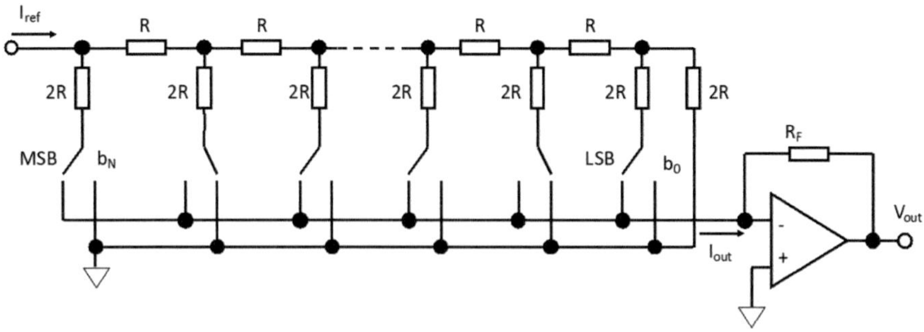

Figure 3.8 R-2R resistor ladder topology DAC for *N*-bit (current-mode).

example with 4-bit). The R-2R ladder employs an R-2R cell for each bit plus a termination resistor, thus reducing the total number of resistors to about 3*N (for example, for a 4-bit 12-resistor R). A ground-connected resistance, 2R, is also added to this architecture, so that the resistance seen from that node is equal to R.

R-2R ladder can generate either a voltage or a current as output. In Figure 3.8, a current is generated and then converted in voltage for direct use by a trans-resistance amplifier. In this circuit, the relationship between the analog output and digital input can be calculated as the superposition of terms that are the successive division of V_{ref} by 2 [2]:

$$V_{\text{out}} = V_{\text{ref}}\left(\frac{R_F}{R}\right)\sum_{i=1}^{N}\frac{b_i}{2^i} \tag{3.7}$$

where N is the bit number. The main disadvantage of R-2R topology is that the input-output characteristic is not intrinsically monotonic, while such a feature is guaranteed only by Kelvin dividers (and their variants).

Regardless of the chosen topology, resistors-DAC can be easily implemented with active (n-well, diffusions, etc.) or passive layers (i.e., polysilicon). We will see in a while that the latter is very convenient for rad-hard designs.

Capacitive-divider DAC

Another widely used approach in DAC design makes use of a capacitive network. This class of DAC is based on replacing the input capacitor of a gain amplifier with an array of programmable capacitors with binary-weighted size. In Figure 3.9, an example of this type of DAC with 4-bit is reported, where Φ_{rst} is a reset phase (all bottom plates are switched to ground) that

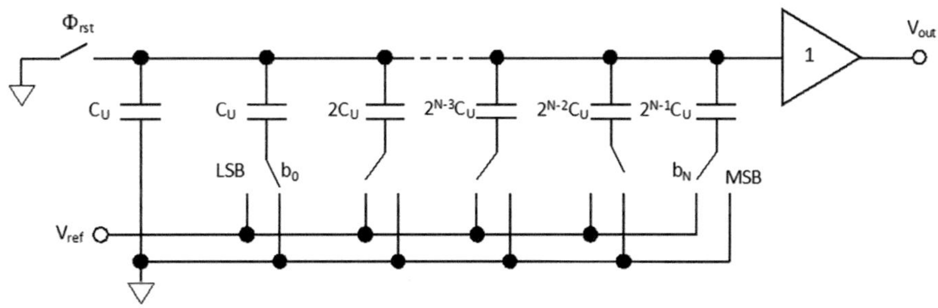

Figure 3.9 Capacitive-divider DAC (N-bit).

precedes the conversion phase itself. The analogue output voltage for a binary weighted array of capacitors is then given by:

$$V_{out} = V_{ref} \sum_{i=0}^{N-1} b_i 2^{i-N} \tag{3.8}$$

The performance of these types of circuits is strongly impacted by the amplifier's finite gain, noise, and offset voltage. Another limitation comes from the presence of parasitic capacitors associated with the unity elements of the cap array. If the parasitic capacitances are dependent on the output voltage, it can cause harmonic distortion. The analysis of mismatch errors and linearity indicates that cons exceed pros for these DACs.

Usually, capacitors are implemented using a "sandwich" of metal and insulator layers, such as in MIM (metal-insulator-metal), which demonstrated to be insensitive to dose, as well.

Current-steering DAC

Another way to design a DAC is to use current generators as a "brick" element. This category includes current-switching topologies with binary-weighted current sources, unary current sources, or mixed methods (called segmented), which use different selection strategies to improve linearity. The principle on which this type of architecture is based is very similar to that of resistor string DACs. The only difference is that instead of using resistors to create different voltage values, current generators are used to create different currents. Then, the analog output voltage is obtained by feeding the output current to a load resistor R_L. Generally, these types of converters are used in high-speed applications (up to GS/s), since they manipulate currents instead of voltages. Figure 3.10 reports two examples of this kind of DAC.

In Figure 3.10(a), an N-bit current-steering DAC deviates M out of 2^N-1 unity current generators toward the output node under the command of the N-bit input digital world. This is called unary weighted control (also called thermometer decoded DAC). In a binary weighted DAC (Figure 3.10(b)), 2^i unity elements are connected in parallel and switched together by the i-th bit. The output current is given by:

$$I_{out} = I_{unity} \left(2^0 b_0 + 2^1 b_1 + 2^2 b_2 + \ldots + 2^i b_i + \ldots + 2^{N-1} b_{N-1}\right) \tag{3.9}$$

Even in this type of architecture, mismatch errors are present. In particular, it is necessary to account for the shunt resistance across each current source and the finite on-resistance of the switches.

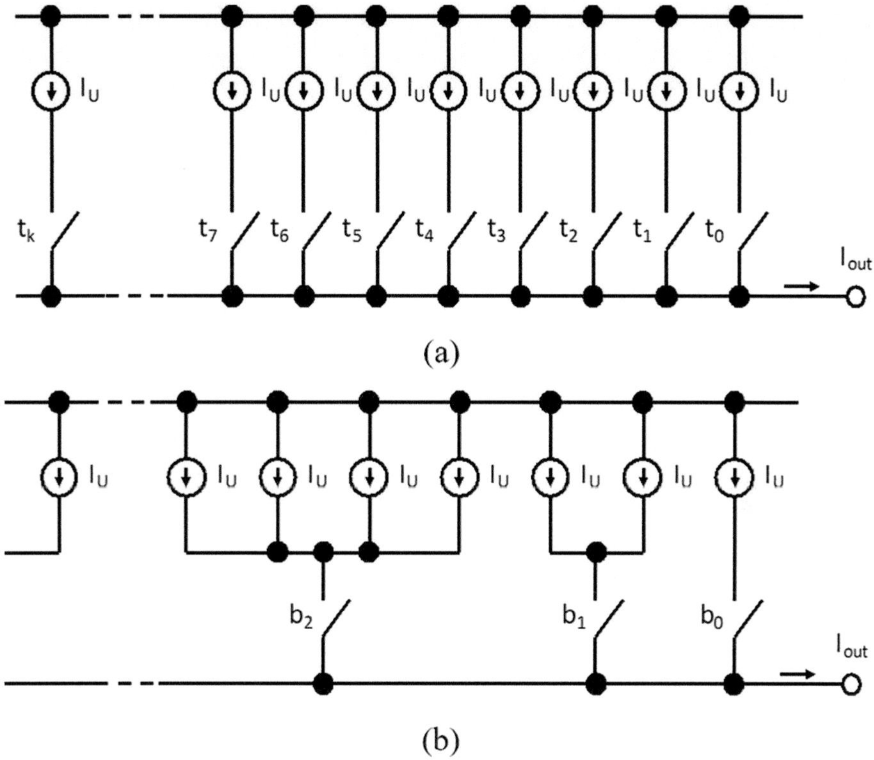

Figure 3.10 Current-steering DAC: unary weighted control (a) and binary weighted control (b).

Both current-steering topologies (a) and (b) show advantages and disadvantages. For a binary weighted DAC, virtually no decoding logic is required; however, it is not strictly monotonic. On the other side, unary DAC is intrinsically monotonic, and with minimum glitches distortion. However, each unity current source requires a logic control signal. For this reason, they have studied digital-to-analog converters, called segmented, that convert a portion of the bits using unary, while the remaining bits are converted using binary weighted [4], [5].

3.3 Rad-hard Digital-to-Analog Converter

The rad-hard digital-to-analog converter represents the interface between a robust digital circuitry and an external "analog" world. Main performance can span from a few kHz to hundreds of MHz for the sampling rate, and from a few to 14-bits as resolution. In order to fulfill these targets, the candidate architectures are different since they shall address different electrical requirements. In addition to this, the DAC shall be resilient to radiation, and this requirement forces toward a specific architectural solution. We have seen above that in the literature, there are several DAC architectures, each addressing different resolution-speed requirements. As said, the most popular are:

- Type 1 – Resistive (resistors, switches, and logic ports);
- Type 2 – Capacitive (capacitors, switches, and logic ports);
- Type 3 – Current-steering (current generators, switches, and logic ports).

In this chapter, we will not analyze in detail other rad-hard topologies, such as sigma-delta based [6], [7], since they address more specific applications and performance ranges. From the electrical point of view, Type 1 is based on a resistive divider (also called a Kelvin divider). It is simple, monotonic, but has several control lines, a high parasitic load at buffer input, and is very variable with code (because of the large number of switches), and poor matching for high bit count. Type 2 is based on the so-called charge sharing technique, already described. Type 3 is suitable for very high-speed, but differential (and consequently integral) non-linearity is strongly dependent on the matching of the binary weighted currents, and so suitable averaging techniques are mandatory, especially under radiation. Moreover, it suffers from high power consumption.

All these architectures shall be assessed with respect to radiation since they exhibit different behaviors as a response. Current-steering is very sensitive to total dose, having the accuracy based on matching between MOS current generators (both unary or binary weighted), and even a little gradient in the variation in the threshold voltages can degrade linearity. On the other side, capacitive DAC, being made up of capacitors and switches, is subject to the hit of high-energy particles, which can alter the charge "content" of the array.

Resistive represents the best compromise. Indeed, DAC resistors in polysilicon are essentially insensitive to radiation, while switches suffer from leakage currents induced by TID, mitigated by MOS length greater than minimum and ELT shapes. In case of a high-energy event, a transient glitch can be observed at the DAC output, but it subsides with the time-constant of the DAC (usually a time of a few nanoseconds) and is reabsorbed, so it must be assessed on a case-by-case basis whether it disturbs or not.

As said, the key point of resistive architectures is that they allow us to employ as a material the silicide polysilicon, which has been demonstrated

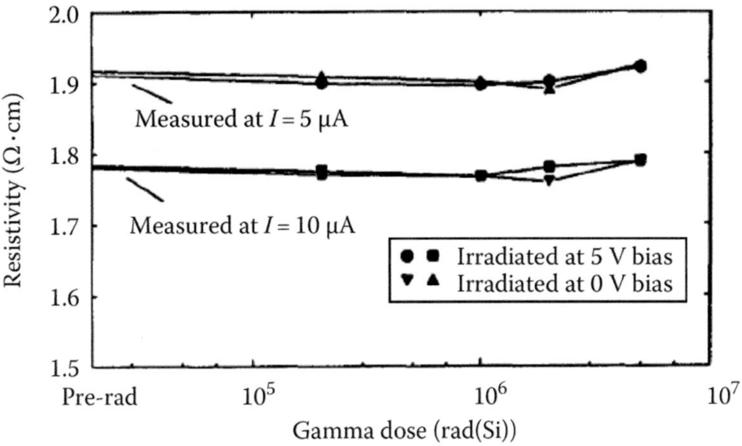

Figure 3.11 Resistivity changes in polysilicon resistors up to 1 Mrad (Si) [8].

to be insensitive to dose [8]-[9]. Please find in Figure 3.11 the behavior of polysilicon resistors.

Several tests have been done in the past on commercial components, showing to what extent TID and SEE can affect device performance [10]-[14]. It is worth noting that the main part of commercial components is radiation tolerant or radiation hard only up to some tens of krad (Si). Generally, they are based on some kind of resistive architecture, confirming the goodness of such an approach. An example of the level of resiliency reached by commercial components in terms of TID and SEL is reported in the following Table 3.1. It has been mainly derived by Ref. [10], [11] and does not pretend to be exhaustive (TID value is referred either to overcoming of specifications limits or to functional failure).

For other converters (such as AD5766 and AD5370), only SEL data is available. From the table, it can be noted that most of the commercial DACs immune to SEL and with good TID performance have been implemented in bipolar or expensive BiCMOS processes. Moreover, bipolar or BiCMOS DACs have a current-steering plus thin-film resistors or pure current-steering topology.

The world of Academia started to work on this topic in order to study solutions to rad-hard DAC issues [15]-[20], as well. Some significant examples will be shown below, the first one being based on resistors, the second and the third on current steering. Further, some words will be spent on the sigma-delta approach [6], [7] without going into details. To the author's knowledge, there are no examples of research ongoing on a capacitor-based

Table 3.1 Commercial rad-hard DAC components [10]-[14].

Part	Manufacturing	Feature	Process	Topology	TID	SEL LET$_{th}$ (MeV*mg/cm^2)
AD9740	Analog Devices	10-bit	CMOS	Current-steering	60 krad (Si)	Low LET
AD667	Analog Devices	12-bit	Bipolar	Current-steering switches and thin-film resistors	75–100 krad (Si)	Immune
AD565	Analog Devices	12-bit	Bipolar	Current-steering switches and thin-film resistors	100 krad (Si)	Immune
HS9-565ARH	Intersil	12-bit	Rad-hard bipolar	R-2R	60 krad (Si)	Immune
DAC121S101QML	Texas Instruments	12-bit	CMOS	Resistors string	100 krad (Si)	120
DAC5675	Texas Instruments	14-bit	BiCMOS	Current-steering	100 krad (Si)	98
DAC5675A	Texas Instruments	14-bit	BiCMOS	Current-steering	150 krad (Si)	75
AD768	Analog Devices	16-bit	ABCMOS	Current-steering switches and thin-film resistors	50–100 krad (Si)	75

rad-hard DAC. Finally, the next paragraph will describe a resistive matrix DAC proposed and developed by the authors.

Resistor-string rad-hard DAC

Starting from the conventional digital-to-analog converters shown in previous sections, different techniques have been applied at circuit, layout, and process levels to make them resilient to dose. The DAC proposed in Ref. [15] presents the implementation of a four-channel 12-bit DAC with high-voltage operation and radiation-tolerant in a 0.5-μm Bi-CMOS technology to achieve the functionality at high-voltage (+/−15 V) and across a wide-temperature range (−55 to 125 °C). The canonical R-2R ladder resistor networks are adopted as a starting point to provide the necessary resistors matching. Any additional digital calibration or laser-trimming is needed to get the resolution. The resistor network is decoupled from the output by means of an inverting buffer. The DAC can satisfy the circuit specifications for low orbit satellite (LEO) application, so as to tolerate single event effects and to fully support total ionizing dose of 10 krad (Si), and with minimum degradation up to 20 krad (Si). The simplified R-2R resistor network is shown in Figure 3.12.

In the R-2R ladder, the linearity of the R-2R ladder is related to the matching of the resistive networks. Generally, the R-2R structure is implemented with thin-film laser-trimmed resistors. Furthermore, the traditional layout of a resistive array usually focuses on the matching accuracy between MSB resistors but does not consider the matching between LSB and MSB resistors, and the matching between LSB resistors themselves. In this work, an unconventional layout technique considers not only the matching between MSB and LSB resistors but also the matching between LSB resistors, therefore significantly improving the DAC precision and linearity. The resistors of the R-2R ladder are split into a different number of resistors with the same size, which can minimize the mismatches between devices and the relevant parasitic resistance and capacitance. Similarly, the resistors in the R-2R are

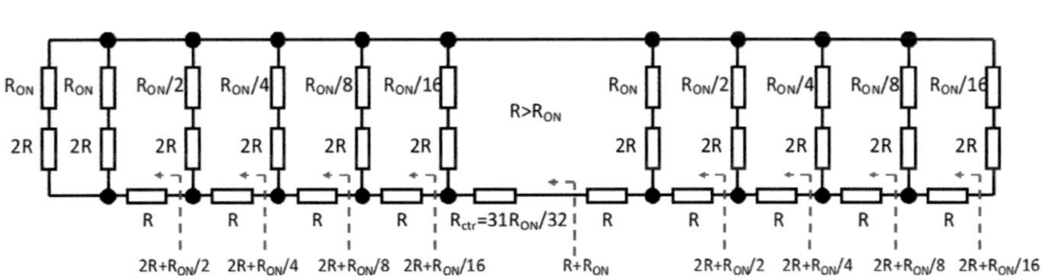

Figure 3.12 Resistor network [15].

symmetrically aligned with only a single metal interconnection trace to minimize the effects of the parasitic resistances of the interconnects. The common centroid layout is not only adopted globally in the whole array, but also in every local region, so that the resistor mismatch caused by stress and temperature gradients is greatly reduced. Moreover, the connections between the entire resistive arrays are everywhere realized with the same metal (Metal1) without vias and multi-layer interconnections, so that the system mismatch caused by them is significantly decreased.

As for radiation, high-precision polysilicon resistors are chosen, since they can meet the requirements of both high voltage operations and insensitivity to total dose. In this design, the analog block adopts a high-voltage (HV) design, and the digital block contains both high-voltage and low-voltage designs. In the analog block, the length of the HV MOS transistors is larger than 3 μm, which can ensure single-event tolerance of the device. Moreover, the tolerance to inter-device leakage is improved by means of guard rings: a triple ring of N+, P+, and N-Well isolates PMOS, while a double ring of P+ and N-Well isolates NMOS.

Redundancy is also a common mitigation strategy to achieve system safety and reliability. In digital circuits, triple modular redundancy (TMR) design is adopted to improve the ability to resist single-event upset (SEU). However, it occupies more area and sacrifices speed, so that TMR is used only in the crucial sub-blocks.

The operational amplifier used as a buffer is a simple two-stage amplifier. For the radiation-hardened design, provisions are taken to improve the input stage immunity by adopting complementary input and a differential structure. A high DC opamp gain (greater than 110 dB) and huge gain-bandwidth (greater than 1 MHz) are enough to compensate for the expected threshold fluctuations.

Given that the electrical performance is typical of this class of circuits (DNL and INL are <1 LSB in all temperature cases), the behavior of the output voltage under total dose of ^{60}Co is optimum till 10 krad (Si), while it starts to degrade around 20 krad (Si) (Figure 3.13).

As concerns SEL, experiments under radiation of $^{35}Cl^{14+}$ ions (13.3 MeV cm^2/mg) and $^{74}Ge^{20+}$ ions (37.5 MeV cm^2/mg) sources demonstrated that separation and guard-rings between PMOS and NMOS in excess of three allows to avoid any latch-up.

One more rad-hard R-2R DAC is described in the Ref. [18]. It has been implemented in a 0.5-um SiGe BiCMOS technology, which features wide temperature capabilities and high-speed. A segmented R-2R ladder architecture has been selected for its simplicity, low-power and less active component

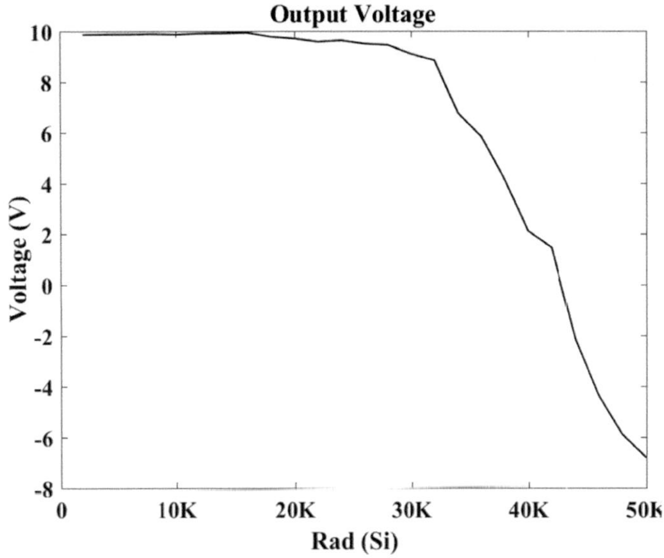

Figure 3.13 Measured output voltage results of the 12-bit DAC [15].

use. It features 8-bit segmented in 5 LSBs binary and 3 MSBs with thermometer codes (Figure 3.14).

Reference current generator and current differential-to-single output converter are implemented with an accurate MOS cascode current mirror starting from a UWT (ultra-wide-temperature) bandgap reference. In order to make the generators resilient MOS length higher than minimum lithographic has been used, while RHBD provisions have been taken against SEL: active N+ (P+) guard rings are added around pMOS (nMOS); n-well and p-substrate

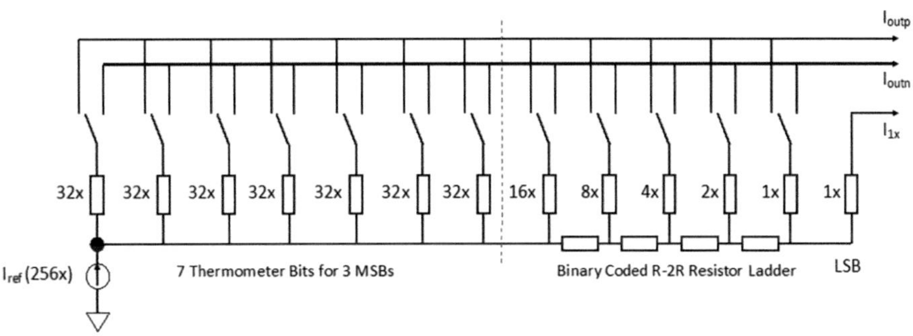

Figure 3.14 8-bit DAC with segmented architecture [18].

contacts are widely used; deep trench rings are added for isolation between pMOS and nMOS; n-well and n+ source/drain are kept as farther as possible (with some penalties in area). The measured DNL/INL at room temperature is well below +–0.5 LSB, while the variability over −120 to 180 °C is less than +–1 LSB. Unfortunately, any measurements under radiation are given in the Reference. Finally, Ref. [19] employs a similar architecture, but unfortunately does not provide any measurement results under radiation.

Current-steering rad-hard DAC

A first example of rad-hard current-steering DAC can be found in [16]. Here, besides general electrical requirements (resolution 12-bit, data-rate 80 MS/s), the device shall satisfy radiation tolerance and operation over a cryogenic temperature range (–180 to 120 °C) required for lunar applications. The technology used is a 0.5-um BiCMOS SiGe technology, which is an expensive and not common process, but can guarantee immunity to radiation for SiGe bipolar devices [8] (employed, for example, in bandgap reference circuits).

This current-steering DAC is implemented with a segmented architecture, made up of three sections: 6 MSB + 4 NSB + 2 LSB (Figure 3.15) (MSB most-significant-bit, NSB next-most-significant-bits, LSB least-significant-bit). The block diagram includes a thermometer decoder, current switch logic

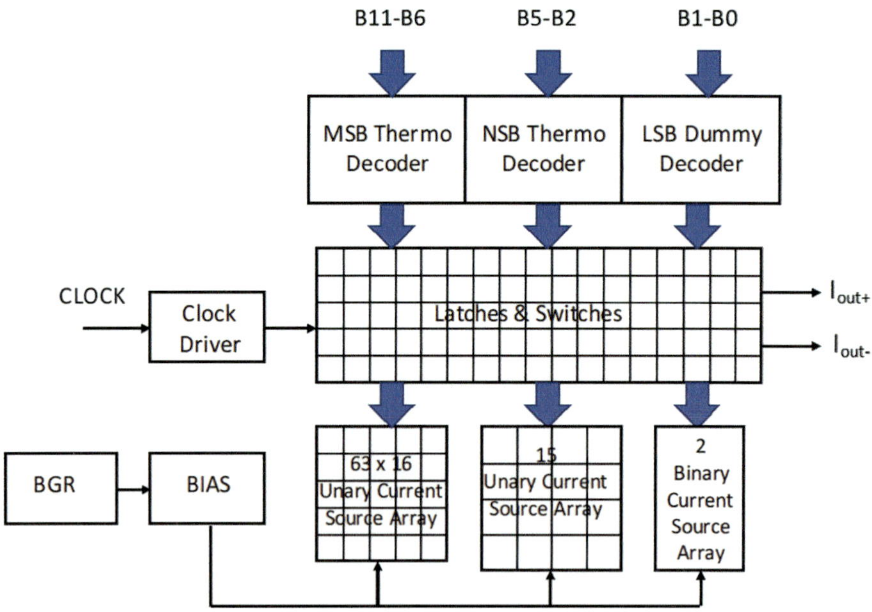

Figure 3.15 Block diagram of the cryogenic DAC in [16].

array, segmented current source array, clock driver, and bandgap voltage reference. Partial segmentation is adopted and implemented by combining some segmentation in the MSBs with a binary weighting for the less significant bits, which can achieve good accuracy and dynamic performance (INL, DNL, glitches, etc.) with an acceptable chip area and circuit complexity. The segmentation in 6 MSB + 4 NSB + 2 LSB has been the result of the deep analysis within the optimal range to keep the best balance between minimizing the circuit area of thermometer decoders and optimizing the DAC static and dynamic performance.

While the switches and relevant control circuit have a conventional topology, the bandgap has been designed to support a wide temperature range (−180 to +120 °C). The BGR utilizes uniquely SiGe bipolar transistors to generate a temperature-independent voltage source.

As for radiation tolerance, the focus of the design has been toward total dose, because it causes the most significant damage and degradation of performance in such current-steering DACs, thus limiting the operation lifetime. Indeed, in particular, total dose changes the threshold voltage of MOS transistors. This problem is especially severe for the current-steering DAC, where the output currents are summed up from the precise unit current sources. The radiation causes significant inaccuracy in the unit currents, which inevitably leads to the degradation of static performance, such as INL and DNL, and dynamic performance, such as SFDR and ENOB. The provision taken was to use large W/L transistors, which can partially mitigate the threshold shift and cumulative charge. Moreover, as said above, the bandgap is built up with SiGe bipolar transistors available in the technology, which are insensitive to the dose and offer wide temperature capabilities. Also, the layout has been carefully done, by using dummy transistors and a large area to improve resiliency to dose. Moreover, extended guard rings are used to provide isolation between current sources and reduce inter-device leakages.

Several measurements are reported in the paper. The output current varies by around 100 ppm/°C over the temperature range, confirming the validity of the bandgap designed. The measured SFDR at reduced clock rate (25 MHz) is −54 dBc at −180 °C, while its degradation at 300 krad (Si) is only 2 dB, and this applies quite similarly over the entire temperature range. This demonstrates that the distortion coming from inaccuracy of unit current sources caused by total dose is modest (even if the starting point, −54 dBc, is not a quite high value for 12-bit converters and may be the indication of intrinsic inaccuracy also at room temperature). As for power consumption, at room temperature it exhibits an increase of >15% after the exposure up to 300 krad (Si), and >25% at −180 °C (Figure 3.16). These results make the

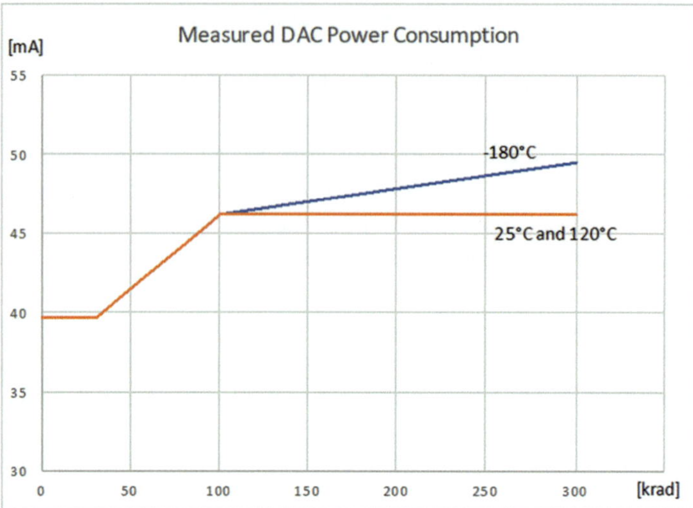

Figure 3.16 Measured power consumption vs. dose rate [16].

proposed DAC suitable for a wide temperature range in radiation-tolerant applications.

A second version of a rad-hard current-steering DAC is presented in [17]. In this chapter, the 12-bit DAC is built up with 2^{12} unit-current cells (Figure 3.17), and the authors analyze the mismatch induced in current cells

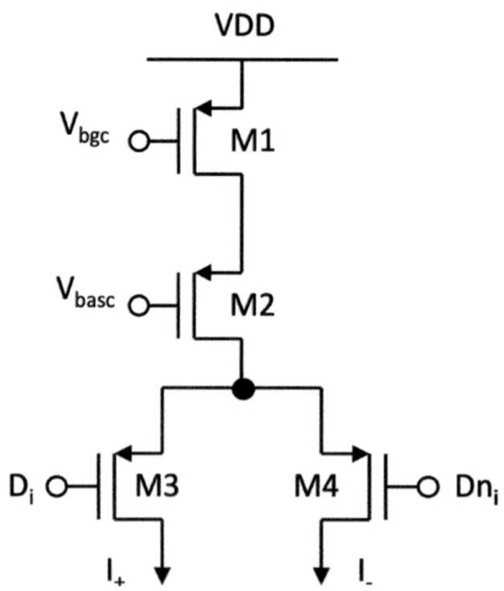

Figure 3.17 Cascode unit-current cell [17].

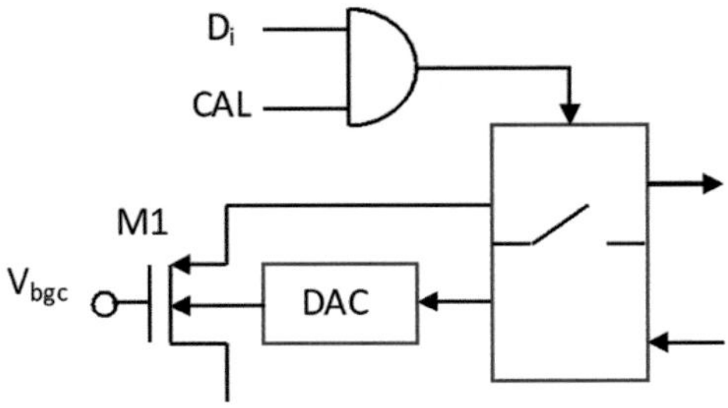

Figure 3.18 Unit-current cell with embedded body-control DAC [17].

by a total dose (using a Gaussian distribution for threshold variations). M1 came out as the most sensitive element to dose.

Using radiation parameters taken from literature and a 65-nm CMOS technology, the evaluated degradation in SNR is from 74 dB (ideal case) to 41.3 dB after 100 Mrad (Si) dose. To overcome this limitation, a technique is proposed that minimizes the mismatch between the unit-current cells. It is accomplished by adjusting the body voltage of the current source transistor M1. During calibration, a simple V_T extractor plus an ADC is used to read threshold voltage shift and apply a correction to the body voltage of M1 in each unit-current cell by means of a resistive-DAC embedded in the cell itself (Figure 3.18).

Since the same calibration circuitry is used for minimizing the mismatch of all unit-current cells, its degradation due to radiation does not have an impact. The technique has been verified by mixed simulation with a Spice-like tool and Matlab®, demonstrating that calibration can recover up to 13 dB in SNR after 100 Mrad (Si) dose.

3.4 New Topologies for Rad-hard DACs

Till now, we have explored the existing solutions that have been proposed to overcome the sensitivity of digital-to-analog converters to radiation and which deal with an RHBD approach at circuit and physical layout levels. Besides making a rad-hard design, new architectures can also be a help, even if less explored. One proposed topology is the sigma-delta DAC. In synthesis, a sigma-delta DAC can be seen as the analog-to-digital conversion in reverse order, where all primary functions of the sigma-delta modulator and digital filtering are the same (see Chapter 4). This DAC solution exhibits the same advantages as its ADC counterpart. Thanks to the intrinsic large oversampling ratio, the requirements on the reconstruction filter are significantly

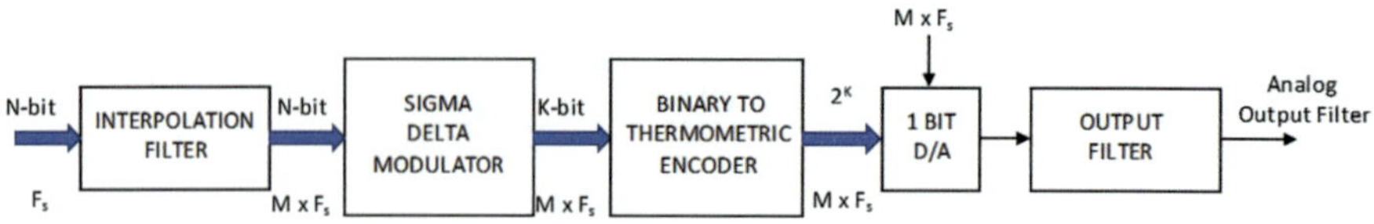

Figure 3.19 Sigma-delta DAC block diagram.

relaxed. Because of their high digital "content" they are also suitable to be used in complex mixed-signal SoCs. The simplified block diagram for a sigma-delta DAC is shown in Figure 3.19.

The N-bit input digital word is fed to a digital interpolation filter, which increases the sampling rate (by a $2f_{bw}$*OSR, where f_{bw} is the input signal bandwidth and OSR is the oversampling ratio, see Chapter 4). It can be thought of as an "under-decimation." The all-digital sigma-delta modulator block performs the noise-shaping function and reduces the number of bits to k, for example, to one bit. Typically, it has a low-order loop-filter, which low-passes the input and high-passes the quantization noise. Then, the k-bit digital stream is converted into analog by means of a k-bit DAC. The number of DAC bits (k) is usually a trade-off between DAC complexity (matching between elements is needed), clock rates, and input bandwidth. Finally, a reconstruction output analog filter removes quantization noise present at high-frequencies caused by digital approximation and rejects images above the Nyquist band.

The work proposed in [6] represents almost an implementation of the block diagram shown in Figure 3.18 with a 130-nm CMOS technology. The interpolator is composed of four stages and realizes a programmable OSR ×128 or ×256. For this design, a third-order sigma-delta modulator has been chosen with a 5-bit quantizer. Being a multi-bit resolution, a dynamic element matching technique is necessary to ensure enough DAC accuracy. In particular, a DWA (data weighted averaging) technique has been applied to the 32 pMOS current sources. Then, these last are summed up and fed to the output by means of a proper circuitry that includes a bandgap block, an RC low-pass filter, and a low-noise operational amplifier (Figure 3.20).

As concerns rad-hard provisions, the DAC has been designed to tolerate a linear energy transfer (LET) greater than 70 MeV mg/cm^2 and a TID greater than 100 krad (Si). The device should also be robust to SEU and SEL, which can disturb the functionality or destroy the converter's performance. The 130-nm CMOS technology used includes radiation-tolerant standard cell libraries. They have been validated through measurements with heavy ions and

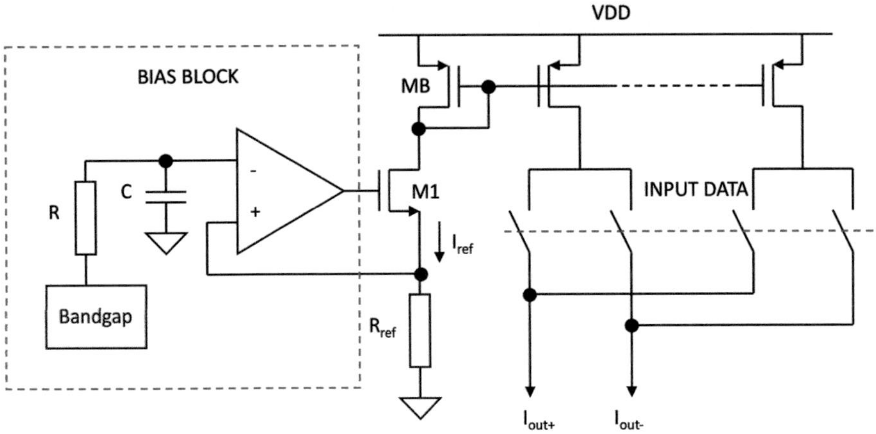

Figure 3.20 Topology of the analog section of the sigma-delta DAC [6].

protons without having a latch-up or a hard fail up to 120 MeV mg/cm^2. An intrinsic radiation hardness of the process was also observed against gamma rays. The irradiated test resulted in no degradation up to 100 krad (Si) [20]. The digital part is implemented using the "robust" cells from the mentioned library, including latches and flip-flops with a bigger area and a higher critical charge tolerance. Furthermore, to ensure that the basic decision nodes of the DAC are SEU-tolerant, TMR is used.

As for layout, to improve immunity to the latch-up, the complete digital section is placed inside a deep N-well, also known as deep N-well isolation (NISO).

For the analog section, the nMOS devices are inside a deep N-well and are surrounded by p+ guard-rings. The increment of the W/L ratio of the transistors increases the intrinsic capacitance, and further additional capacitances are added to the most sensitive nodes. The separation between the p+ diffusion in the well and the n+ in the substrate is increased, and additional substrate and well contacts have been placed wherever possible.

Measurements before radiation indicated a DNL and INL of 0.5 LSB, and a SINAD of 108 dB for an input sinewave of 100 Hz. After a dose of 100 krad (Si), no change in the value of the monitored parameters is observed, that is, output current capabilities and analog and digital power consumption.

A second version of the same design has been implemented with a 150-nm process [7]. Besides already used radiation-hardening provisions, the ELT shape has been adopted for nMOS.

3.5 3MS/s 10-bit and 12-bit Poly-Resistor Rad-hard DACs

Starting from the consideration that the resistive-string solution seems the most promising in terms of simplicity, consumption, and resiliency, the authors developed and verified a 3 MS/s DAC having a 10-bit embedded interface in microcontrollers [21]. The technology used was a 130-nm CMOS. The target performances of the prototypes are total dose immunity up to 300 krad (Si), resiliency to latch-up up to 60 MeV*mg/cm^2, and SEU of 10 MeV*mg/cm^2 for the digital controlling section. Our target architecture has been the so-called "X-Y addressing scheme with shunt resistor" or "intermeshed architecture" [22], [23] which is an improved variant of the folded resistor string, discussed in the previous paragraph. The simplified schematic, including X-Y selection switches, is shown in Figure 3.21.

The coarse interval of voltages is defined by the main resistor string (RCx), while the sub-strings (RFy), placed in parallel to the coarse, define the taps for the fine voltages. RCx are selected with switches driven by the

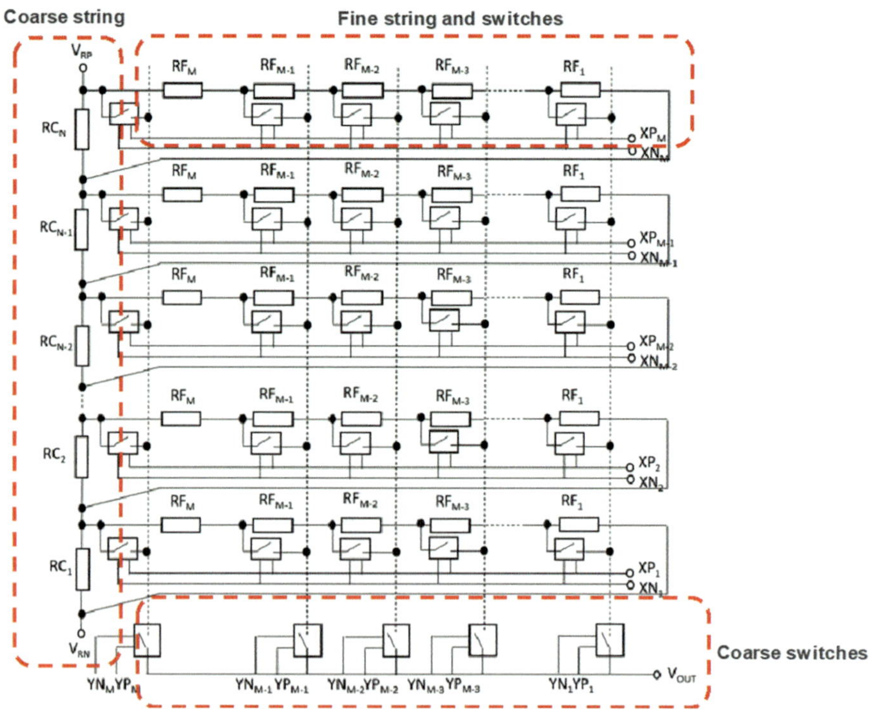

Figure 3.21 Schematic of the X-Y resistor matrix DAC [21].

5-MSB, while RFy are selected by means of switches driven by 5-LSB. In this design, all resistors are the same size.

This architecture offers the following advantages with respect to the conventional Kelvin divider:

- Reduced number of bit control lines;
- Time-constant of the resistor matrix is kept low without using low-resistance unity elements.

The latter point is very important since it allows for consuming less power across the string: the current across it is V_{RP}-V_{RN}/R_{TOT}, where R_{TOT} is the total resistance seen between references. Moreover, this topology allows the use of resistors with a suitable value, which does not degrade matching. Indeed, the linearity performance of the matrix (and so of the DAC) is strongly related to the matching between resistors. The higher the area and perimeter of the polysilicon resistor, the better the matching. To complete the DAC functionalities, decoders, control logic, and an output buffer have been designed. The block diagram is shown in Figure 3.22.

Suitable X (for RCx) and Y (for RFy) decoders transform the input 10-bit code into X-Y address of the matrix, selecting the proper voltage tap that gives the analog voltage corresponding to the input bit code (such as selecting a box in a chessboard). Suitable registers synchronize the input

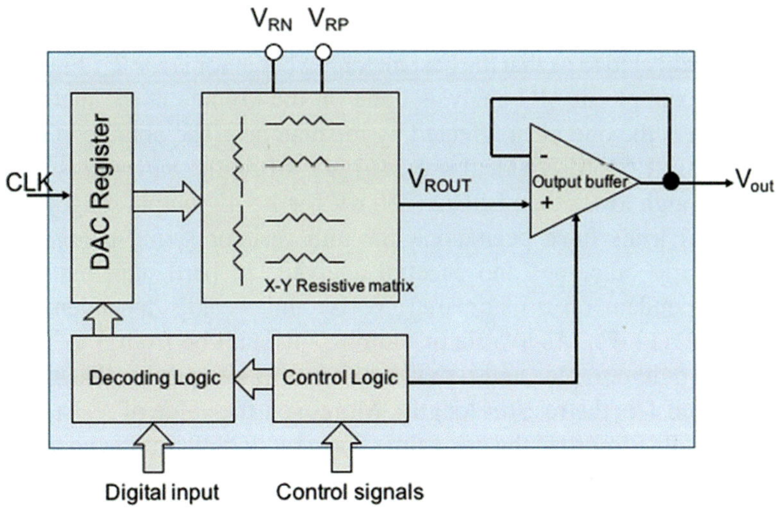

Figure 3.22 Complete block diagram of the full DAC.

bit, while a control logic oversees DAC operations and implements Reset and Hold functions. A closed-loop buffer decouples the output load from the resistor matrix, thus improving speed performance.

One of the core activities in the design is the sizing of the unit resistor. As mentioned above, its physical characteristics, size, and area have an impact on converter performance. Once the best matrix topology for the intended use is selected, resistors material is the next critical choice. The adopted criteria have been:

- Resistor value
- Resistor matching data
- Resiliency to radiation

Resistors based on "active" layers are discarded because of their sensitivity to charges. In the literature, resistors based on a polysilicon layer have been demonstrated that are insensitive to radiation, as reported in [8] and verified also in the ADC presented in [24], so we employed them. Three kinds of polysilicon resistors were available in the process:

- R_{high} (high-ohmic poly-Si resistor)
- R_{ppd} (medium resistance poly-Si resistor)
- R_{sil} (salicided poly-Si resistor)

Their sheet resistances are typically 1300, 250, and 7 Ohm/square, respectively. Statistical simulations (MC, Montecarlo) for process and mismatch variations indicated to us that the best matching behavior is for R_{sil}. Figure 3.23 reports an example of MC analysis done on the voltage at the matrix mid-code, which is the one most affected by mismatches. The obtained Gaussian mean value is 1.64665 V (ideal = 1.6467 V), while the variance is 310 uV, which is enough lower than LSB/2 (750 uV for 1.5 V_{pp} output range).

Several trials have been done for unit resistor sizing using statistical Montecarlo outcomes and spectral analysis. In particular, matching is strongly dependent on resistor area (W*L) and weakly dependent on the perimeter (2*(*L*+*W*)). As a "rule of thumbs," *W* shall be from 3 to 10 times the minimum lithographic and *L* shall be from 5 to 10 squares (*W* is the resistor width, and *L* is the resistor length). Moreover, the value of resistors shall be made small to improve the array time constant, which for five coarse unity resistors R_{U} is approximately given by:

$$t_{\text{DAC}} = \left(2^5 / 2 * R_{\text{U}} + 2 * R_{\text{on}}\right) * \left(C_{\text{p}} + C_{\text{L}}\right) \tag{3.10}$$

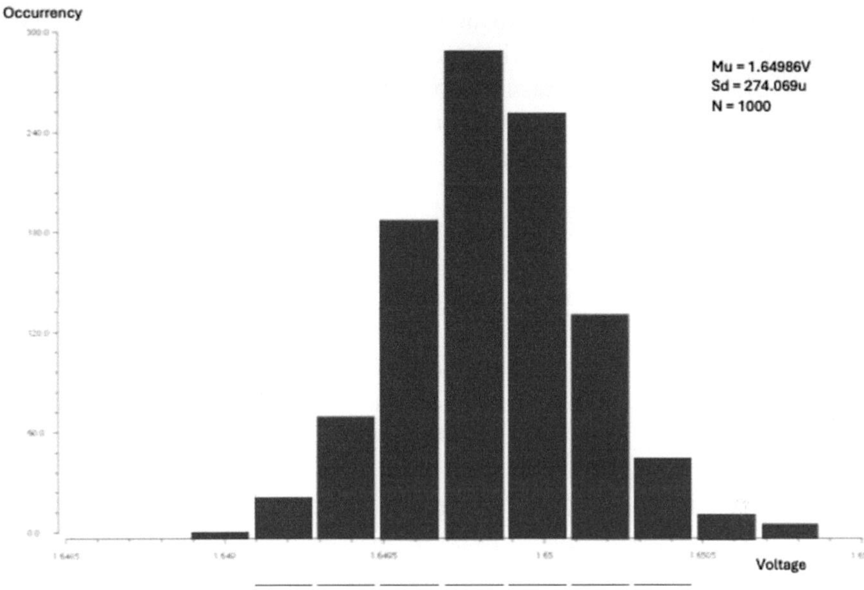

Figure 3.23 Montecarlo simulations for process and mismatches variation (500 Runs) – RCx = RFy = 61 ohm [22].

where R_{on} is switches on-resistance, C_p is the switches parasitic capacitance, and C_L is the load capacitance of the resistor matrix, i.e., the input buffer capacitance. However, power consumption is proportional to $1/R_{TOT}$, so the resistor value cannot be too low.

By finding the best compromise between area occupation, statistical performance, power consumption, and linearity, RCx = RFy = 61 ohm. The matrix is composed of 2^5 RCx resistors and by 2^5 fine strings, each of them implemented with 2^5 RFy.

Also, for the sizing of switches sizes, some opposing "rules of thumbs" have been followed:

- For speed: switches on-resistance shall be low (i.e., *W*/*L* ratio higher than minimum)
- For parasitic capacitance, switches shall be small
- For resiliency, *L* shall be higher than the minimum lithographic or ELT shape shall be adopted [25].

Switches on-resistance is made more constant vs. source/drain voltages using a complementary pair, thus mitigating non-linearity in R_{on} and so in the network time-constant. Complementary CMOS switches help in clock-feed-through reduction, as well. Matrix time-constant, t_{DAC}, is proportional to both switch on-resistance and parasitic capacitance. Larger switches have lower on-resistance (proportional to *L/W*) but higher parasitic capacitance and area, so that a trade-off has been found.

The layout of the resistor matrix plays an important role in determining DAC performance. Indeed, the resistance of a real implementation differs from the nominal value, R_U, because of processing effects. In particular, the resistor is affected by an absolute error and a relative mismatch, usually with a gradient in one direction. Let us consider, for example, a straight string with unity elements R_U. It can be calculated that the resistor INL has a parabolic shape. If the resistive divider is folded around its mid-point, the maximum INL is reduced by a factor of 4 [2] (see Figure 3.24, an example with four resistors).

We followed the folding principle by making the layout of a matrix where rows from 1-to-$2^5/2$ are placed in the left matrix, rows from $2^5/2+1$-to-2^5 in the right matrix. Moreover, having a resistive matrix layout with similar length and width is convenient with respect to gradient processing (Figure 3.25). Red arrows indicate the direction of folding for coarse and fine resistor strings. In addition, fine resistors are intermeshed.

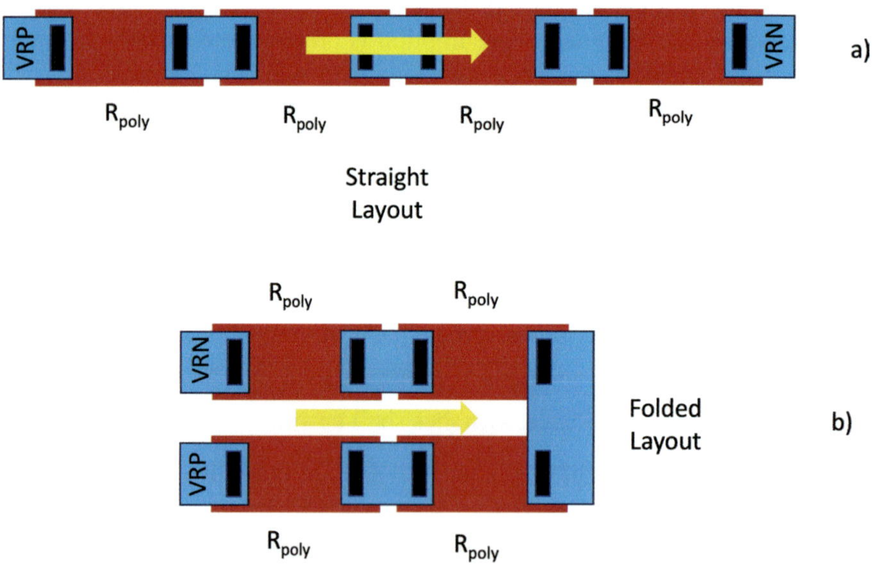

Figure 3.24 Straight-layout (a) and folded-layout (b) of a resistor string [22].

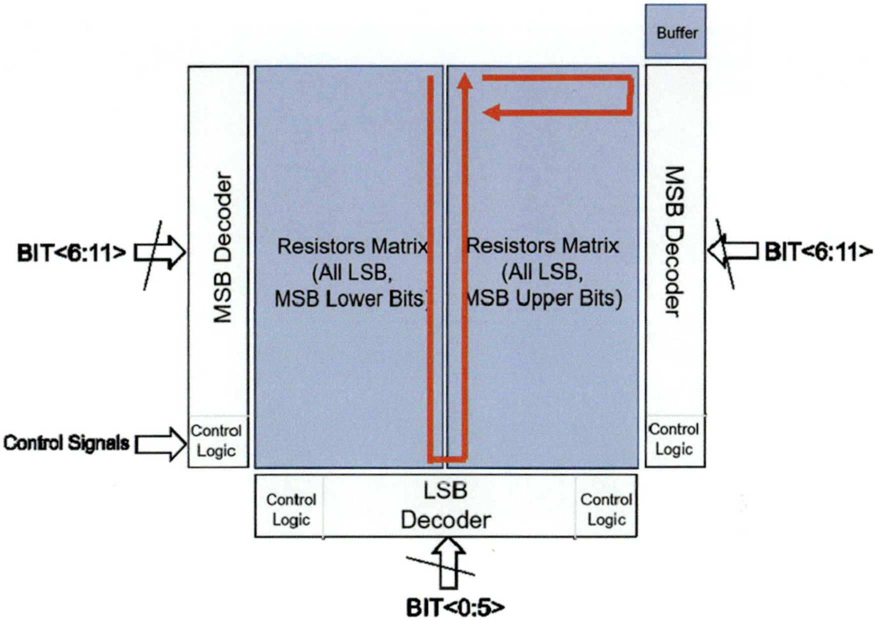

Figure 3.25 Floor plan of the 10-bit DAC [22].

To mitigate total dose leakage as much as possible, the switches layout has been designed with an ELT (edge-less transistor) shape. An example of a switch physical layout is shown in Figure 3.26. It can be noted that the annular shapes of the MOS gate to reduce intra-device leakage is used, while supply (VDD) and substrate (GND) guard-rings around the transistors overcome inter-device leakage and counteracts also possible SEL effects.

For the digital part, SEU can be an issue in the input digital latches, which freeze the input bit. The use of robust latches (where additional capacitances are added to the most sensitive nodes) belonging to a custom rad-hard library helps to avoid any bit-flip.

The cells library, implemented in similar technologies, has been characterized in the past and demonstrated to be resilient to a TID of 300 krad (Si), SEU of 10 MeV*mg/cm^2, and SEL of 60 MeV*mg/cm^2. The DAC digital section (control logic and decoders) has been designed using such a library. In the overall DAC layout, EGR (extended guard-ring) is placed wherever possible to prevent SEL.

The design of the buffer required a bigger effort to avoid analog performance degradation, as well. Figure 3.27 shows the schematic of the main opamp. It is a symmetrical opamp with a Miller compensation [26].

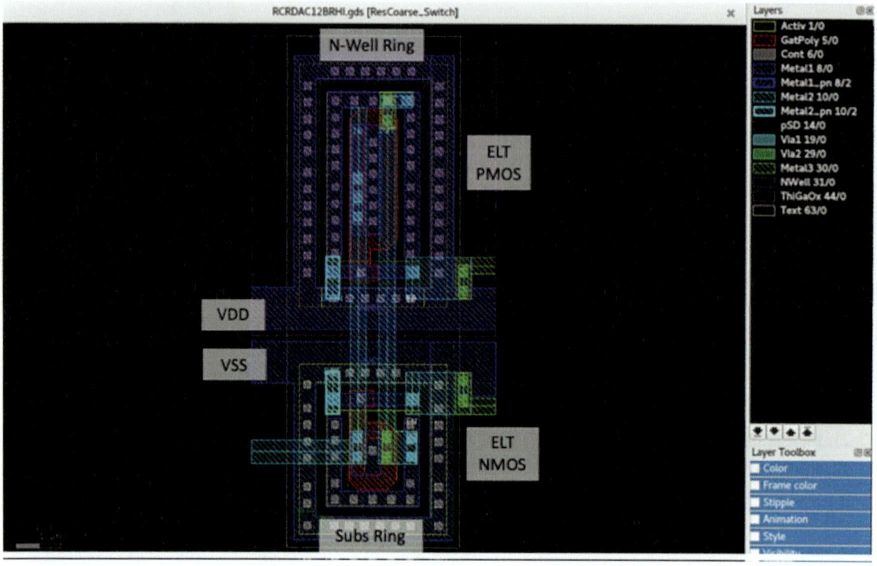

Figure 3.26 Layout of the switch used as a coarse and fine selector in the DAC matrix.

This architecture has been chosen since it is simple while guaranteeing enough gain.

Because of sensitivity to dose, the more complex the opamp, the more probable is the performance degradation. Rad-hard provisions in the opamp

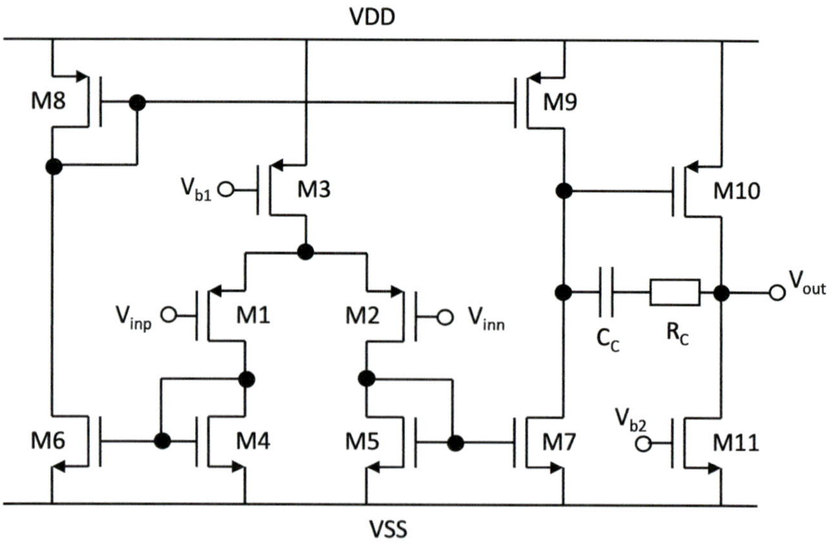

Figure 3.27 Schematic of the opamp used as DAC output buffer.

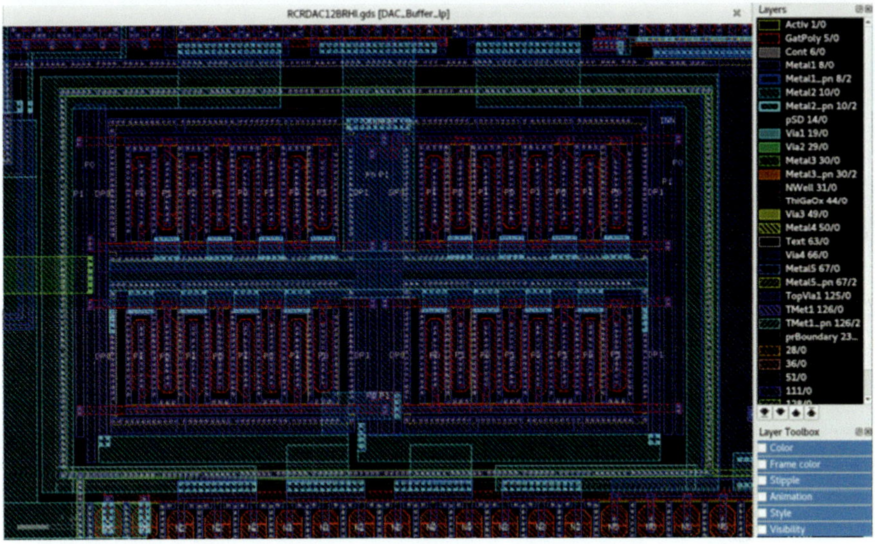

Figure 3.28 Layout of the opamp input differential stage with common-centroid ELT.

are mainly demanded by layout techniques. Moreover, the open-loop gain has been kept high to allow for a degradation under radiation. All transistors used in the buffer circuit have been sized to consider the radiation tests reported in literature (*W* and *L* not minimum), and applying the ELT shape. The opamp offset is mainly determined by the matching of the input pair and directly reflects as a shift of the DAC transfer curve. The transistor's threshold voltage mismatch is not influenced by the radiation induced threshold voltage shift as assessed in [27]. To achieve the physical matching of the input stage, we applied a variant of the common centroid topology of ELT MOS. For example, the MOS of the input pair (M1, M2) has a $W = 160$ um each. They have been split into 16 ELT, each of them placed in a "common centroid" way (Figure 3.28). The same technique was applied to the active load and bias circuit.

The layout of the DACs is shown in Figure 3.29. The size is 0.9 mm × 0.7 mm (0.63 mm^2) for the 10-bit and 1.6 mm × 1.4 mm (2.24 mm^2) for the 12-bit.

As can be seen from the above pictures, the proposed DAC architecture is modular and can be easily up-scaled. The authors designed the 12-bit version of the DAC starting from the 10-bit. The "extension" has been achieved by means:

- Increase coarse and fine decoders capability (from 5 + 5 to 6 + 6)
- Increase coarse string (from 32 to 64 poly resistors)

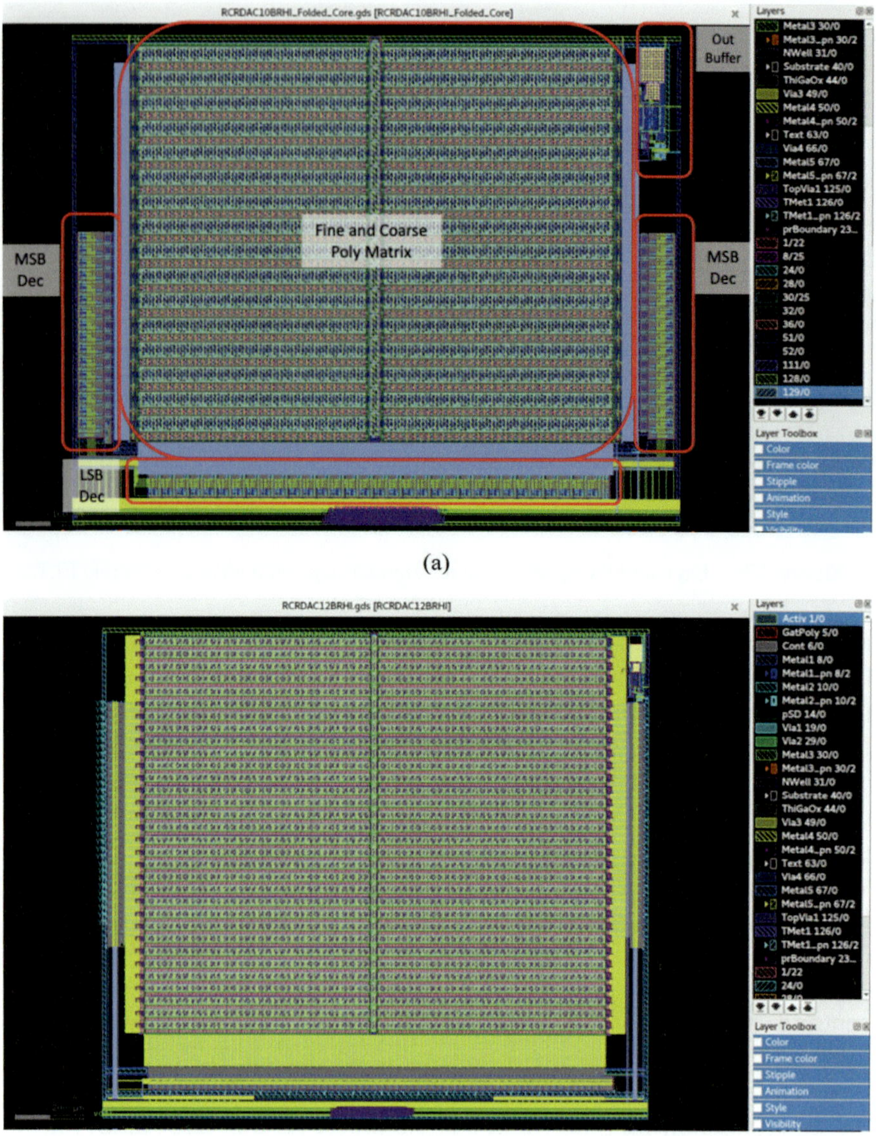

Figure 3.29 Layout of the 10-bit (a) and 12-bit (b) DAC macros [22].

- Increase fine strings (from 32 to 64 poly resistors each)
- Same floor plan

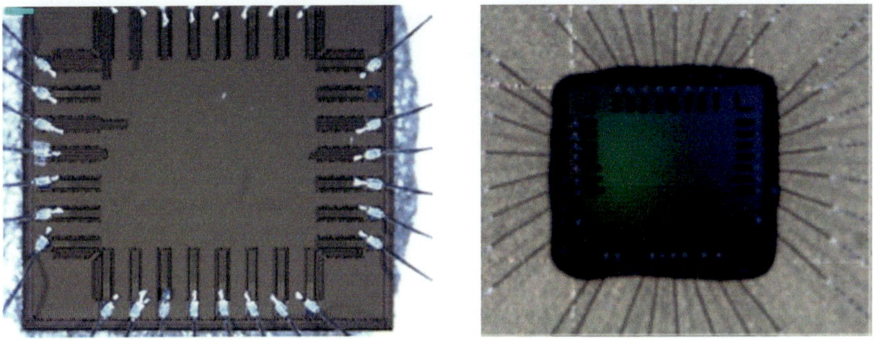

(a) (b)

Figure 3.30 Chip photo for 10-bit (a) and 12-bit DAC (b).

The resistor-string DAC has been integrated in a standard 3.3 V 130 nm CMOS process (IHP SG13S). Figure 3.30 shows the microphoto of the two DACs. The overall die size of the 10-bit chip is 1.7 mm × 1.5 mm (2.55 mm^2), including the pad-frame, while the overall chip area for the 12-bit is 2.5 mm × 2.25 mm (5.62 mm^2).

The component (12-bit) assembled in a 32-pin CERDIP (ceramic dual-in-line package) is shown in Figure 3.31.

3.5.1 In-lab measurements

Both 10-bit and 12-bit prototypes have been preliminarily characterized in terms of static and dynamic performance. The output voltage range has been fixed at 1.5 V. The DNL and INL (presented in Figures 3.32 and 3.33) have

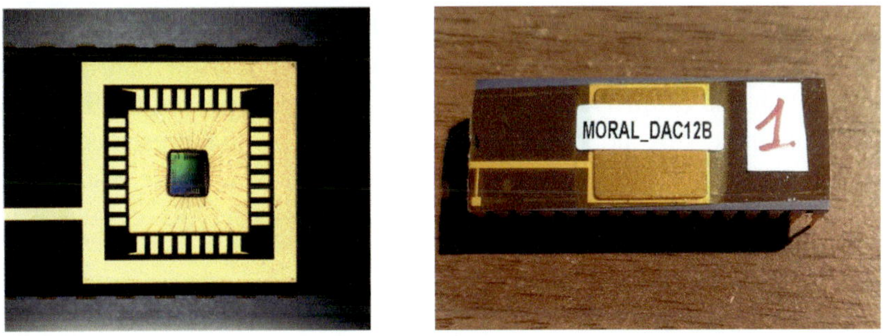

(a) (b)

Figure 3.31 Photo of the 12-bit DAC: (a) internal; (b) external.

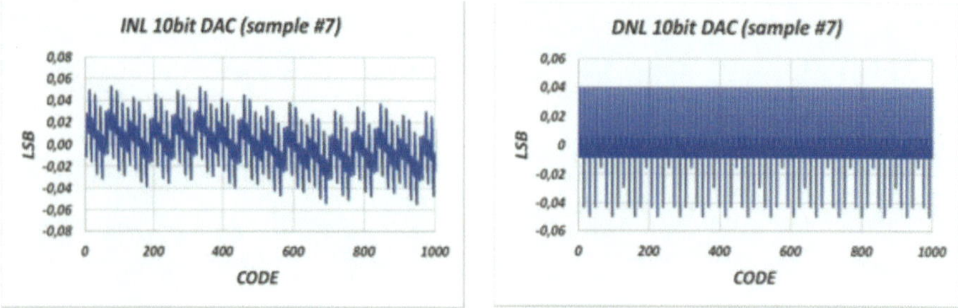

(a) (b)

Figure 3.32 Measured DNL (a) and INL (b) of the 10-bit DAC.

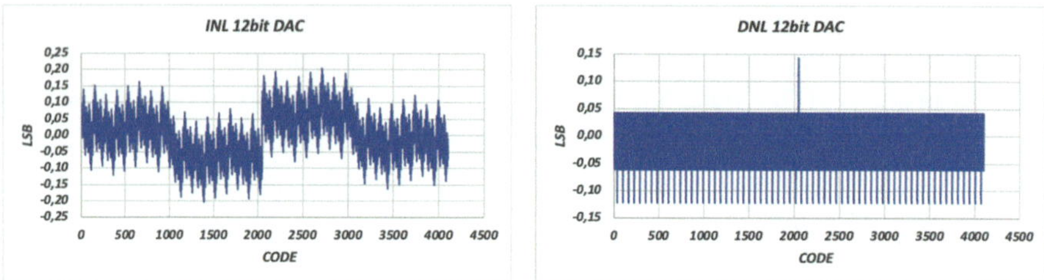

(a) **(b)**

Figure 3.33 Measured DNL (a) and INL (b) of the 12-bit DAC.

been measured by applying the so-called "major carrier method" [28], which can be used for such monotonic converters and allows for saving time in testing.

The measured 10-bit DNL is less than ±0.1 LSB, while INL is less than ±0.1 LSB, as well. While the measured DNL is less than ±0.15 LSB, the INL is less than ±0.2 LSB for the 12-bit. These excellent results have been obtained thanks to the accurate layout of the resistor matrix (folded approach) and an extensive try-extract-simulate-repeat procedure. The validity of the up-scaling approach is demonstrated by the presented results, where linearity is maintained while bit count increases. These results make the converter very suitable to be used in on-board satellite DC- or quasi-static applications.

The dynamic performance has been measured by applying a sinewave and using a clock of 1 MS/s and then a clock of 3 MS/s. ENOB measurements at 1 MS/s vs. input signal frequency are shown in Figure 3.34. ENOB accounts for noise and total harmonic distortion (THD).

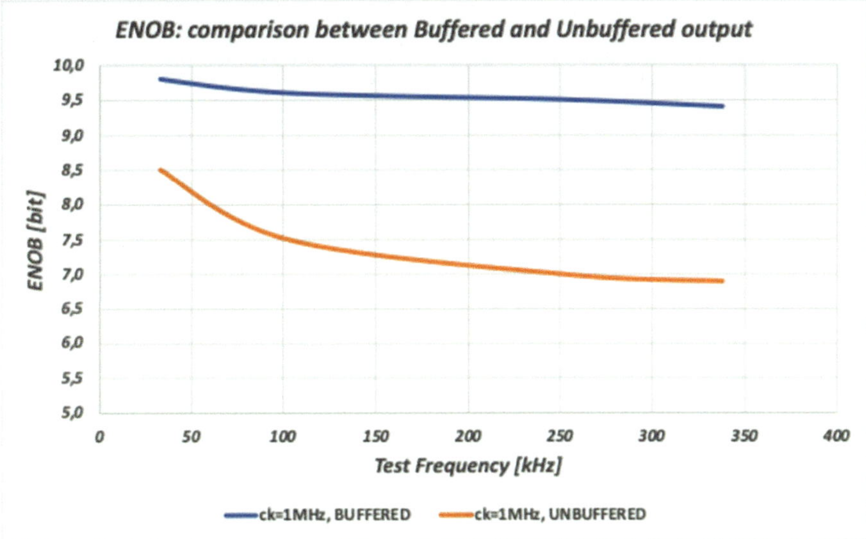

Figure 3.34 Measured ENOB vs. input frequency of the 10-bit DAC with 1 MS/s clock.

For the 10-bit, both buffered output and unbuffered output (at the matrix output node) are shown. ENOB is around 9.8 bits for buffered, while unbuffered is worse because of the time-constant determined by capacitive load and the overall matrix resistance. In the latter case, at 1 MS/s and 33 kHz input frequency, ENOB degrades by more than one bit. In Figure 3.35, the spectrum of the buffered output when a 33 kHz 1.5 V_{pp} sinewave with a clock of 1 MS/s is applied is shown.

Same measurements have been performed on the 12-bit DAC. Figure 3.36 shows the spectrum of the buffered output when a 33 kHz 1.5 V_{pp} sinewave with a clock of 1 MS/s is applied, while Figure 3.37 reports the spectrum when a 100 kHz 1.5 V_{pp} sinewave with a clock of 3 MS/s is used. The ENOB is higher than 10.8 bits for 1 MS/s and 9.2 bits at 3 MS/s. At 3 MS/s, the ENOB degrades approximately to one bit and a half. The DAC remains operative at 2 V_{pp}, but the SFDR is degraded by 5 dB.

Intermodulation (IMD) performance (IP3) has been measured, as well. A pair of tones equal to 217.4 kHz + 232.6 kHz is applied with a clock of 1 MS/s, and the IMD is 65 dBc (Figure 3.38). IMD decreases at 56 dBc when the tones are equal to 652.174 kHz + 697.674 kHz at a clock of 3 MS/s.

Core power consumptions (string current excluded) from a 3.3 V supply are in Table 3.2.

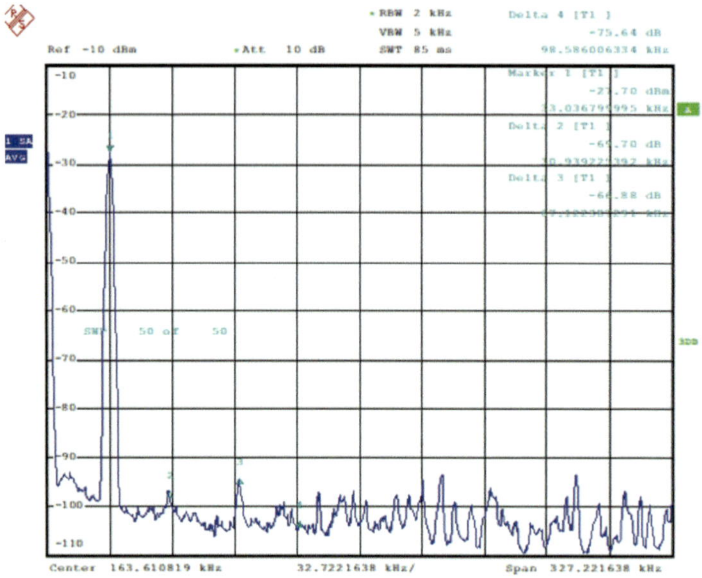

Figure 3.35 10-bit DAC measured spectrum of the buffered output when a 33 kHz 1.5 V_{pp} sinewave with a clock of 1 MS/s is applied (ENOB = 9.8 bit).

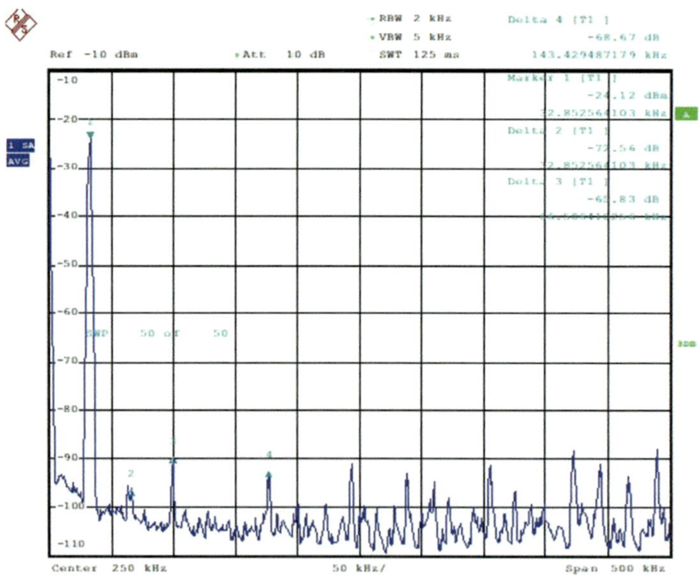

Figure 3.36 12-bit DAC measured spectrum of the buffered output when a 33 kHz 1.5 V_{pp} sinewave with a clock of 1 MS/s is applied (ENOB = 10.8 bit) [22].

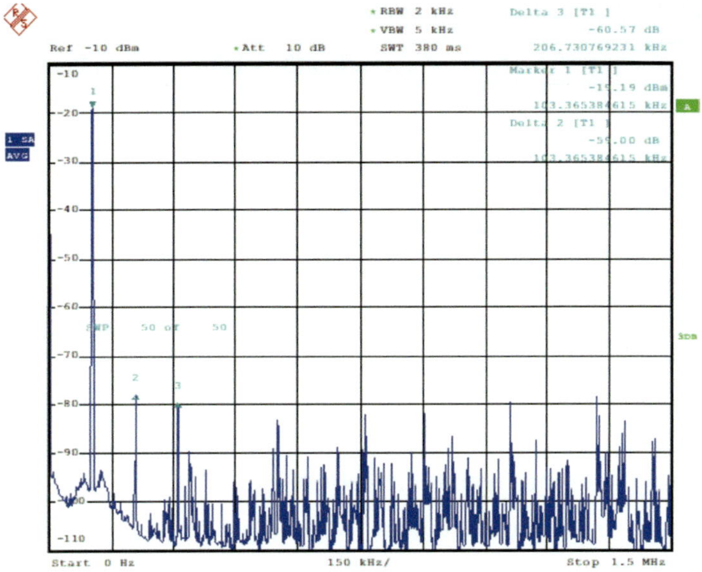

Figure 3.37 12-bit DAC measured spectrum of the buffered output when a 100 kHz 1.5 V_{pp} sinewave with a clock of 3 MS/s is applied (ENOB = 9.2 bit).

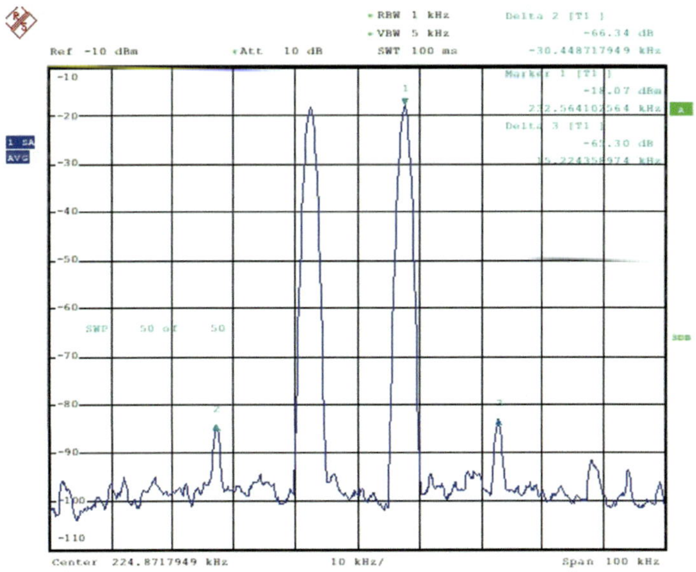

Figure 3.38 12-bit DAC measured IMD (IP3) when the buffered output of two tones with a clock of 1 MS/s is applied (IMD = 65 dBc) [22].

Table 3.2 Core power consumptions (string current excluded) with a 3.3 V supply.

	Normal operation (3 MS/s)	Power-down
10-bit DAC	10 mW	<200 uW
12-bit DAC	10.5 mW	<200 uW

The main part of the consumption (3 mA) is due to the output buffer for both, driving a capacitive load of 10 pF. The small difference in consumption is because the two DACs are produced in two different batches. The string current consumption is approximately 0.39 mA (12-bit) for a 1.5 V output voltage range.

3.5.2 Measurements under radiation

The rad-hard 12-bit DAC has been tested under ^{60}Co at the University of Palermo facility using the RedCat Devices proprietary Leonida Platform. Using a 0.63 rad/s dose rate, during the experiment, the DAC reached up to 500 krad (Si) with no significant degradation in terms of electrical performance.

Due to the extreme environmental conditions, especially the high levels of radiation, it is crucial to develop an appropriate testing methodology. Traditional methods – such as using a test board with the device under test (DUT) mounted and lab equipment to generate signals and record outputs – are not viable. Only the DUT should be subjected to radiation, while measurement tools and auxiliary circuits must remain outside the irradiation zone. As a result, all measurement procedures must be conducted remotely.

The proposed solution is based on splitting the test board into two boards: a motherboard (MB), which hosts all the controls, exposed; and a daughterboard (DB) with the DUT only. This approach is described by means of several examples in Chapter 5. Here, as the motherboard, an off-the-shelf Arduino Due® board plus a custom shield are used. The analog output of the proposed DAC is converted into digital by a very accurate ADC onboard the Arduino Due, and then recorded. In the following, the measurement results obtained under TID with such methodology are presented.

For the TID measurements, we sequentially applied three digital input codes to the DAC (corresponding to 0, mid-scale, and full-scale values) and recorded the resulting output using the ADC on the Arduino Due. Additionally, we measured the static power consumption – primarily attributable to the analog section – using a suitable current sensing circuit.

Figure 3.39 presents the Arduino ADC output code when the DAC is set to code 2027. The absolute median value recorded (2030) does not match – or closely approximate – 2047, because the Arduino ADC exhibits an offset relative to the DAC's range. Specifically, while the DAC's half-scale output voltage is exactly 1.65 V (VDD/2), the ADC's mid-scale input corresponds to approximately 1.665 V, representing a difference of 15 mV (or 18 LSB). By adding 18 to the measured median (2030), we obtain 2048, which closely aligns with the DAC's mid-scale value. We confirmed that, up to a TID of 500 krad (Si), the ADC readings remain within ±1 LSB, which is primarily attributed to measurement uncertainty (as shown in Figure 3.39).

A slightly worse behavior has been measured for 0 and 4095 codes, with an uncertainty of ±2 LSB, which always stays the same up to 500 krad (Si).

Current consumption is monitored using a current sense circuit located on the shield board, which produces a voltage proportional to the supply current passing through it. This voltage is digitized by another Arduino's ADC and recorded by the developed software (i.e., Leonida Platform as described in Chapter 5). Figure 3.40 displays the analog current consumption in milliamperes when a mid-scale code is applied to the DAC input. As shown, aside from some measurement uncertainty – caused, for instance, by the low voltage drop across the current sense resistor and the limited resolution of the

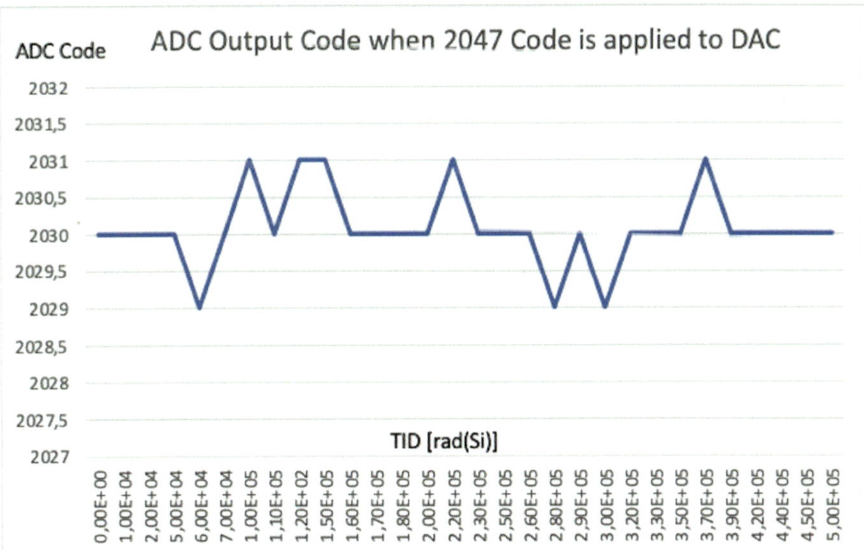

Figure 3.39 Arduino ADC output code (as a digital representation of the proposed 12-bit DAC analog output) as a function of radiation dose [21].

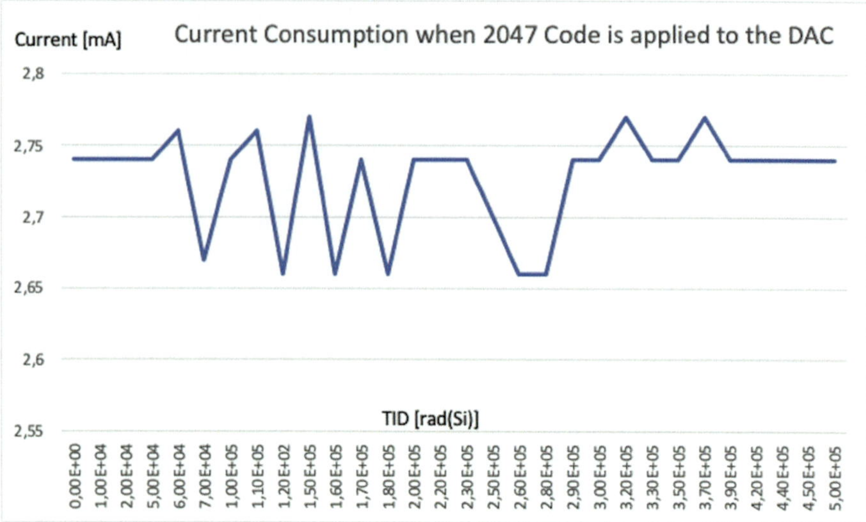

Figure 3.40 12-bit DAC analog current consumption as a function of radiation dose for the half-range DAC [21].

ADC – the current remains nearly constant at around 2.7 mA up to a TID of 500 krad (Si), as illustrated in Figure 3.40.

Moreover, in Figure 3.40, there is a comparison of DAC current consumption as a function of radiation dose for the three codes, 0, 2047, and 4095. As shown, the consumption is higher for the 2047 code, depending on the operation of the DAC output buffer.

Furthermore, Figure 3.41 compares the DAC's current consumption as a function of radiation dose for three input codes: 0, 2047, and 4095. As illustrated, the current consumption is highest for code 2047, which is attributed to the behavior of the DAC's output buffer at mid-scale.

On the other side, digital current consumption is very low and impulsive, with an average value of <300 uA, which does not significantly change with the total dose.

Table 3.3 provides a comparison between the proposed design and previously reported radiation-hardened DACs from the literature. As shown, our DAC demonstrates competitive performance despite being implemented in a standard bulk CMOS process. This is a significant advantage, as bulk CMOS is more cost-effective than BiCMOS and facilitates integration into more complex systems that include substantial digital logic. In terms of radiation tolerance, our design outperforms the others; the DNL/INL metrics are comparable, and the overall area footprint is even smaller.

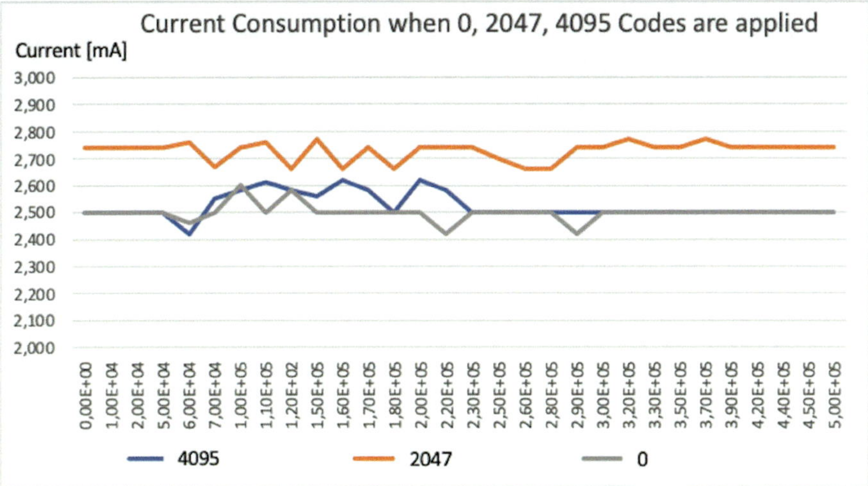

Figure 3.41 12-bit DAC analog current consumption as a function of radiation dose for three codes: 0, 2047, and 4095 [21].

Previous implementations of rad-hard resistive DAC have been reported in [15], [29], and [16]. With respect to the R-2R ladder DAC in [15], we operate at 3.3 V instead of ±15 V, so as to be suitable as an embedded macro in SoC. Moreover, we achieve equivalent INL at a higher speed (1 to 3 MS/s instead of 100 kS/s). The resistive DAC in [29], despite being similar in architecture to ours, has less bit count (8-bit), works at a lower speed (1 MS/s), and is only simulation-proven, not measured in real samples. DAC in [16] has a current-steering architecture and operates at a higher speed (up to 80 MS/s) with similar SFDR (spurious free dynamic range) but is implemented in an expensive 0.5-µm SiGe BiCMOS technology.

3.6 Conclusion

In this chapter, we have analyzed the basic concept of digital-to-analog converters, which is a fundamental block used on the border between the digital and analog worlds and often used as an embedded block in SoC. It is crucial to study how to design these classes of circuits appropriately since their performance affects speed, dynamic range, accuracy, etc. of the interfaces of the circuits or systems where they are inserted. This is particularly true in harsh environment electronics, where, besides the usual causes of degradation (temperature, power supply variations, marching, etc.), resiliency to radiation also plays a crucial role. Thus, we illustrated an overview of the

Table 3.3 Performance comparison with some rad-hard DACs taken from the literature.

	12-bit DAC	**10-bit DAC**	**[15]**	**[18]**	**[16]**	**[20]****	**[17]**
Resolution	12 bits	10 bits	12 bits	8 bits	12 bits	12 bits	10 bits
Technology	130 nm CMOS	130 nm CMOS	0.5 um BiCMOS	0.5 um SiGe BiCMOS	0.5 um SiGe BiCMOS	65 nm CMOS	130 nm CMOS
Data rate	3 MS/s	3 MS/s	100 kS/s	10 MS/s	80 MS/s	NA	15 MS/s
DNL	±0.15 LSB	±0.1 LSB	±0.23 LSB	±0.2 LSB	NA	NA	NA
INL	±0.2 LSB	±0.1 LSB	±0.67 LSB	±0.3 LSB	NA	NA	NA
SFDR	69 dBc	67 dBc	NA	55 dBc	54 dBc	NA	NA
ENOB	9.5 bits	9.2 bits	NA	NA	NA	NA	8.73 bits
TID	500 krad (Si)	500 krad (Si)	20 krad (Si)	NA	300 krad (Si)	100 Mrad (Si)	NA
Architecture	X-Y R Matrix	X-Y R Matrix	R-2R	R-2R	Segmented current steering	Current steering	Current steering
Supply	+3.3 V	+3.3 V	±15 V	+3.3 V	+3.3 V	+1.2 V	+1.2 V
Area (macro)	2.24 mm^2	0.63 mm^2	20 $mm^{2,*}$	0.25 mm^2	6.3 $mm^{2,***}$	NA	NA

*Four channels of the DAC; **only the simulation was provided; ***includes pads.

main design techniques proposed in literature to overcome this issue, with some examples, with the aim of giving the reader useful design insights and an almost complete reference list with the most recent proposals. Moreover, we described rad-hard DACs developed by the authors, based on a resistive approach, and the main results obtained on real measurements.

References

[1] R. Jacob Baker, "CMOS Circuit Design, Layout, And Simulation" R. Jacob Baker Ed.. IEEE Press.

[2] F. Maloberti, "Analog Design for CMOS VLSI Systems", Springer, 2001.

[3] B. Razavi, "Principle of Data Conversion System Design", IEEE Press, 1995.

[4] M. Pelgrom, "Analog-to-digital Conversion", Springer 2016.

[5] Topics on Microelectronics, Course Proceedings, Milan, 25-27 May 2021.

[6] G. Tsiligiannis, et al., "A Very High Resolution DAC at 1kHz for Space Applications", 2010 NASA/ESA Conference on Adaptive Hardware and Systems, USA, Aug. 2010.

[7] M. Karaolis, et al., "High Resolution Radiation Hardened DAC in CMOS-SOI Featuringa Return-to-Zero Matrix", AMICSA 2016, Gothenburg, Sweden, 2016.

[8] G. Cressler and H. Mantooth Ed., "Extreme Environment Electronics", CRC Press, 2013.

[9] C. L. Axness, et al., "Radiation Characteristica of SIPOS and Polysilicon Resistors", IEEE Trans. on Nucl. Science, Vol. 38, No. 6, Dec. 1991, pp. 1365-1369.

[10] F. Irom and S. G. Agarwal, "Compendium of Single-Event Latchup and Total Ionizing Dose Test Results of Commercial Digital to Analog Converters", Proc. 2015 IEEE Radiation Effects Data Workshop (REDW), 2015.

[11] S. E. Stone, et al., "Single Event Effects Characterization of Maxwell 7846B DAC", Proc. 2012 IEEE Radiation Effects Data Workshop, 2012.

[12] A. B. Karakozov, et al. "Single Event Effects and Total Dose Testing of Digital to Analog Converters", Proc. 17th European Conference on Radiation and Its Effects on Components and Systems (RADECS), 2017.

[13] A. C. Daniel, et al., "Radiation Effects Characterization of Commercial Multi-Channel Digital to Analog Converters for Spaceflight Applications", Proc. 2019 IEEE Radiation Effects Data Workshop, 2019.

[14] K. Kruckmeyer, et al., “Analysis of Low Dose Rate Effects on Parasitic Bipolar Structures in CMOS Processes for Mixed-Signal Integrated Circuits”, IEEE Trans. Nucl. Sci., vol. 58, No. 3, June 2011, pp. 1023–1031.

[15] H. Fan, et al., “A 4-Channel 12-Bit High-Voltage Radiation-Hardened Digital-to-Analog Converter for Low Orbit Satellite Applications”, IEEE Trans. on Circuits and Systems–I, Vol. 65, No. 11, Nov. 2018, pp. 3698-3706.

[16] Y. Yao, et al., “A 12-Bit Cryogenic and Radiation-Tolerant Digital-to-Analog Converter for Aerospace Extreme Environment Applications”, IEEE Trans. on Industr. Electr., Vol. 55, No. 7, July 2008, pp. 2810-2818.

[17] A. T. Kunnath, et al., “A Digitally Assisted Radiation Hardened Current Steering DAC”, Proc. 29th International Conference on VLSI Design, 2016, pp. 559-560.

[18] Zhenqi Chen and Fa Foster Dai, “A 3mW 8-Bit Radiation-hardened-by-design DAC for Ultra-wide Temperature Range from -180°C to 120°C”, Proc. 2011 IEEE International Symposium of Circuits and Systems (ISCAS), 2011, pp. 997-1000.

[19] L. Dugoujon, C. Papadas and B.Sinnis, “A 0.13um CMOS Rad-hard proven technology with associated mixed mode circuit design approaches for space applications”, AMICSA 2006.

[20] Kim, D.; Kwon, I.; Kim, C.-H. A 10-bit Radiation-Hardened R-2R Digital-to-Analog Converter for Nuclear Power Plant Applications. In Proceedings of the Transactions of the Korean Nuclear Society Spring Meeting, Jeju, Republic of Korea, 23–24 May 2019.

[21] C. Calligaro, U. Gatti, “A CMOS 10-Bit 3MS/s Rad-hard DAC Based on Resistor String Poly-Matrix”, Proc. MIEL 2021, September 2021.

[22] C. Calligaro, U. Gatti, “A CMOS 12-Bit 3MS/s Rad-hard Digital-to-Analog Converter Based on a High-Linearity Resistor String Poly-Matrix”, Chips 2024, 3(2), 129-152; https://doi.org/10.3390/chips 3020006

[23] M. Pelgrom, et al. “A 50MHz 10-bit CMOS digital-to-analog converter with 75 Ω buffer”, in Jour. of Solid-State Circuits, Vol. 25, Dec. 1990, pp. 1347-1352.

[24] U. Gatti, C. Calligaro, E. Pikhay, Y. Roizin “Radiation-hardening methodologies for Flash ADC”, Analog Integrated Circuits and Signal Processing Journal (Springer), DOI 10.1007/s10470-015-0627-7.

[25] G. Anelli et al., “Radiation tolerant VLSI circuits in standard deep submicron CMOS technologies for the LHC experiments: Practical

design aspects," IEEE Trans. Nucl. Sci., Vol. 46, No. 6, Dec.1999, pp. 1690–1696.

[26] T. Vergine, et al., "A 65 nm Rad-hard Bandgap Voltage Reference", IEEE Trans. Nucl. Sci., vol. 63, No. 3, June 2016, pp. 1762–1767.

[27] J. Veerbeck, et al. "Radiation effects upon the mismatch of identically laid out transistor pairs", IEEE ICMTS Int. Conf. on Microelectronic Test Structures, 2011, pp. 194-197.

[28] Mark Burns and Gordon W. Roberts, "An Introduction to Mixed-Signal IC test and Measure", Oxford University Press 2001.

[29] Aishwarya Harneer Suresh, Corrado Carta, Gunter Fischer, "Design of radiation-tolerant digital-to-analog converter and investigation on analog single event transient effects", AEU - International Journal of Electronics and Communications, Volume 187, December 2024, 155503.

4

Rad-hard Analog-to-Digital Converter (ADC)

Abstract

We have seen in the previous chapter the importance of data converters in electronics systems, both for terrestrial and space applications (or other harsh environments). The analog-to digital converters are not an exception, being widely used in almost every electronic equipment, also on-board satellites.

In this chapter, focused on ADC, we first recall its specifications, which give general information about the features and limits of the static and dynamic operations. Then, we will illustrate the basic principles of ADC and main topologies (flash, pipelined, SAR, etc.). As happened for DAC, starting from known architectures and applying rad-hard design provisions (RHBD), we will describe which are the most promising rad-hard ADC presented in the literature. It is worth noting that the number of research papers dealing with rad-hard ADC is much greater than its DAC counterpart and testifies to the widespread use that is made of these components. Finally, we will also give a quick look at other solutions for specific applications, such as single and dual-slope converters for nuclear experiments. Lastly, a rad-hard analog-to-digital converter recently implemented by the authors is detailed.

4.1 Introduction

As similarly featured by the DAC, the analog-to-digital converter (ADC) operates as an interface between the analog and digital worlds. It converts a continuous amplitude signal belonging to the analog domain into a digital logical expression "0" and "1," with different coding schemes. ADC performance is often a bottleneck for practical applications because of the non-idealities it introduces, such as poor accuracy, speed limitation, distortions, etc. Most terrestrial applications of ADC today can be classified into four huge market segments: (1) data acquisition, (2) precision industrial measurement,

DOI: 10.1201/9788743813286-4

(3) voice-band and audio, and (4) "high speed" (implying high-sampling rates and often a telecom use). A very large percentage of them can be fulfilled by flash, successive-approximation (SAR), sigma-delta (Σ-Δ), and pipelined architectures. A basic understanding of these most popular ADC architectures is a useful supplement to a further in-depth analysis of rad-hard converters that are widely used in space. Indeed, as DAC, the ADC converter represents a fundamental component as a stand-alone device, for example, in telecom. Telecommunication satellites with on-board digital processors require ADCs at the receiver to digitize the IF/baseband carrier information. Mobile missions require digitizing information bandwidths of some tens of MHz, thus necessitating baseband sampling at a rate greater than 100 MS/s. Future broadband missions require digitizing information bandwidths of hundreds of MHz, thus necessitating sampling at a rate greater than 1 GS/s. On the other side, ADCs are critical blocks as embedded macros in complex systems such as microcontrollers used in satellite as housekeeping or for propulsion control, etc. In this case lower speed is needed (in few MHz range or less).

The aim for "space" IC designers is to realize a high-performance radiation-tolerant or even radiation-hard ADC by using standard commercial processes of integration that are inexpensive, with easy maintenance, and widely available in the industry.

4.2 Analog-to-Digital Converters: Basic Principles

The analog-to-digital conversion can be represented as the cascade of four functions (Figure 4.1): continuous time anti-aliasing filtering, sampling, quantization, and data coding.

First, a low-pass filter (also called an anti-aliasing filter) limits the input signal bandwidth so that the sampler does not alias any unwanted signal component into the actual signal band. The sampler transforms a continuous-time signal into its sampled-data discrete-time equivalent. Then, the amplitude of the signal is quantized, i.e., approximated with a set of equally-spaced levels, thus producing a discrete-amplitude signal. Finally, the codifier generates a digital representation of that level, according to different coding schemes, such as unipolar straight binary, or complementary straight binary, etc. [1], [2].

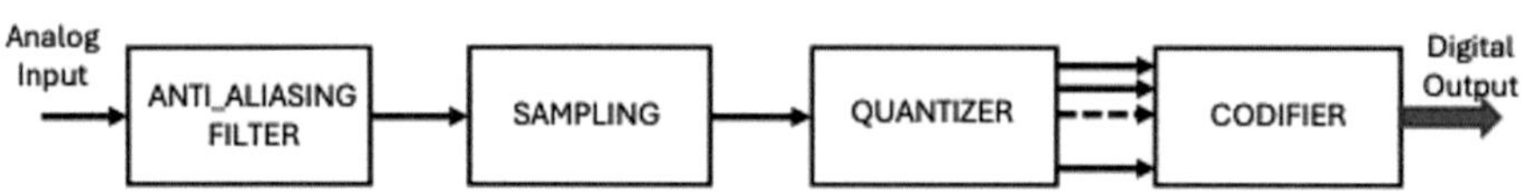

Figure 4.1 Block diagram of the basic functions of an ADC.

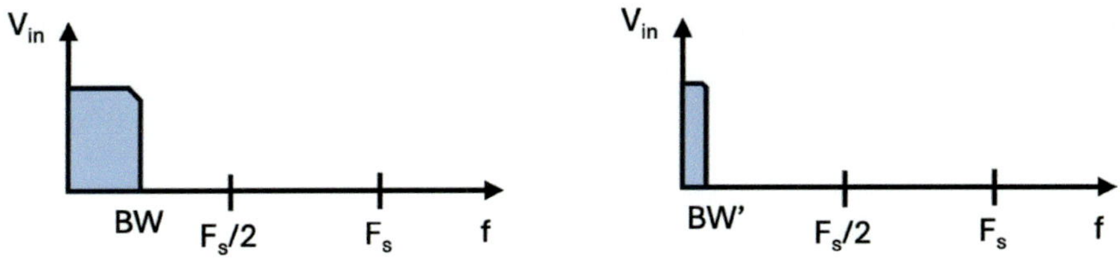

(a) (b)

Figure 4.2 Nyquist-rate (a) and oversampled converters (b).

The analog resolution of an ADC is the smallest analog increment corresponding to a 1LSB code change at the output (LSB defined as $V_{LSB} = V_{ref}/2^N$, N is the bit number and V_{FS} is the full-scale voltage).

The ADC can be divided into two classes depending on the ratio between signal bandwidth (BW) and sampling rate f_s. Nyquist converters are those where the sampling frequency is slightly higher than twice BW (Figure 4.2). In oversampling converters (such as sigma-delta), the sampling frequency is many times the BW's frequency, while output digital filtering is used to remove out-of-band quantization noise.

The ratio between the Nyquist limit and the signal band, $F_s/(2^*\text{BW})$, is called the oversampling ratio (OSR). Converters with a large OSR are called oversampling converters, whereas Nyquist-rate converters have a small OSR, typically less than 8.

For a deeper insight into the characteristics, function, and theory of ADC blocks, the authors refer to focused publications (Refs. [1]-[4]).

Many characteristics define an ADC's performance. Each specification will be discussed before we look at the basic architectures in the next section.

4.2.1 ADC main performance

ADC performance is described by means of some specifications. Here, we indicate only those who differ from that one already described for the DAC. They are [2]:

Static

- Resolution
- Offset
- Gain error
- INL/DNL
- Missing code and Monotonicity

Dynamic

- Sampling rate
- SNR
- ENOB/SINAD/SFDR
- Dynamic range

An ADC generates a digital code that represents the value of an analog input. Since the input signal can take on infinitely many values, but the output is limited to a finite number of digital codes, the resolution of the converter is determined by the number of bits in its output word (for example, 12 bits). An N-bit ADC can therefore represent 2^N distinct levels. The ideal transfer characteristic is shown in Figure 4.3. The smallest input change that produces a variation in the output is $\Delta = V_{FS}/2^N$, where V_{FS} (also called V_{ref}) is the reference full-scale voltage of the converter. The input range of the ADC is defined with respect to this reference. The step size Δ corresponds to the least

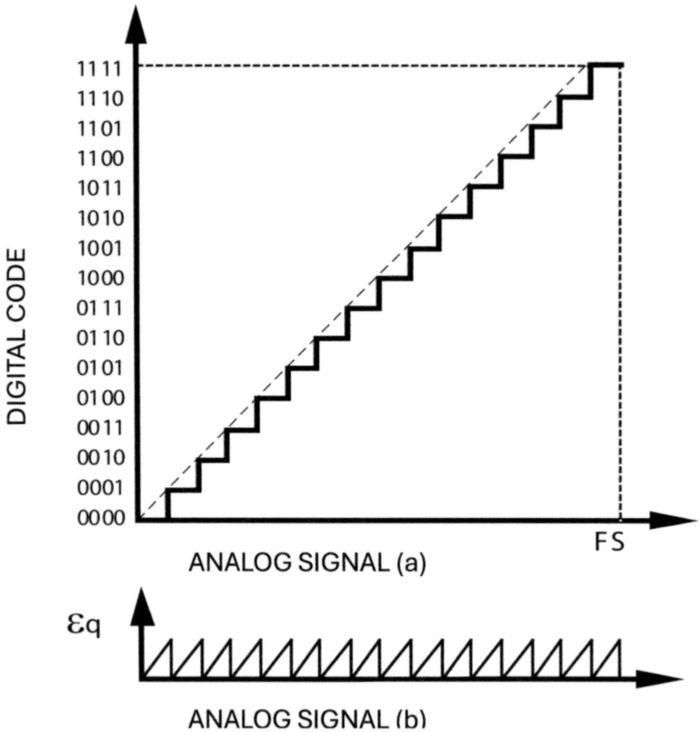

Figure 4.3 Ideal transfer curve of an ADC.

significant bit (LSB) of the digital output. Eq in Figure 4.3b is the quantization error.

The second characteristic is the offset, which is defined as the output bit word value corresponding to $V_{in} = 0$ V and can be determined by the vertical intercept of the straight line through the end point of the ideal input-output ADC transfer curve (Figure 4.4). The offset is expressed in units of LSB.

The third parameter, the gain error (V_{gain}), is defined as the deviation of the slope of line AB from its ideal value, which is usually equal to one (Figure 4.5).

The fourth parameters are the integral non-linearity (INL) and the differential non-linearity (DNL), shown in Figure 4.6(a) and (b), respectively. As for DAC, INL is the maximum deviation of the transfer characteristic from a straight line passing through its endpoints. The difference plot is called the INL profile. DNL is the deviation in the difference between two consecutive code transition points on the analog input axis from the ideal value (1LSB). Both are expressed in units of LSB.

A missing code is, for example, when an increase (or a decrease) of the analog input signal leads to an increase (or decrease) of two LSBs; that is, one digital code is missing. Or we can define that we get a missing code also when DNL = −1.

Monotonicity is when an increase (decrease) in the input signal may result in a step tower lower (higher) digital output code.

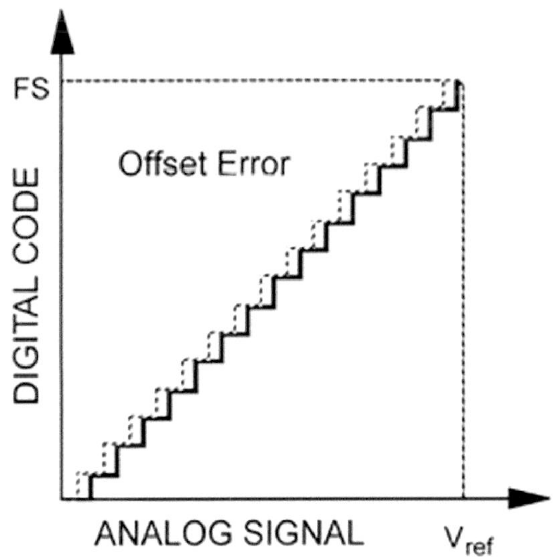

Figure 4.4 ADC transfer curve with offset error.

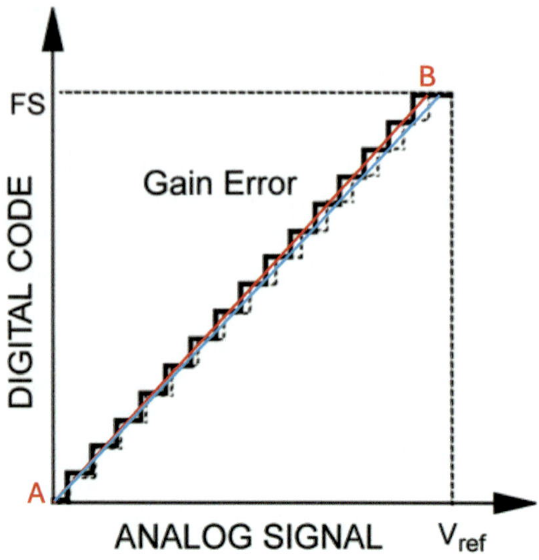

Figure 4.5 ADC transfer curve with gain error.

The highest conversion rate an ADC can achieve depends on the time required to complete one conversion cycle. Beyond any additional processing, this limit is largely dictated by the settling time and bandwidth of the input sampling stage. The analog input bandwidth of an ADC is defined as the frequency at which a full-scale input signal produces an output that is 3 dB lower than its low-frequency value.

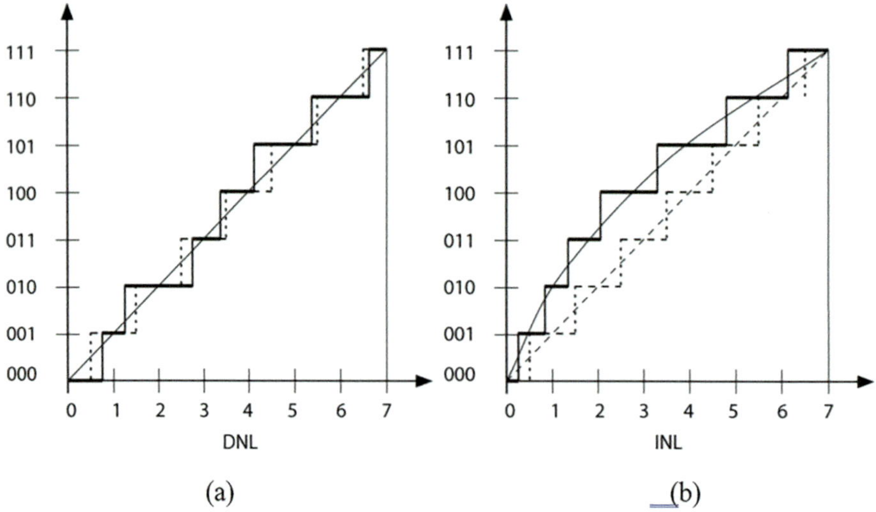

Figure 4.6 ADC DNL (a) and INL (b) [1].

Quantization error for ADC is defined in the same way as in DAC and represents the difference between the original analog input and the digitized output. It is still given by SNR = 6.02*N + 1.76 (dB). In a real converter, the noise power shall also include the noise (thermal, jitter, etc.) of the circuits.

Other dynamic parameters, such as SINAD, ENOB, and SFDR, are common to DAC.

The dynamic range is defined as the ratio of the power of a full-scale sinusoidal input to the power of a sinusoidal input for which SNR = 0 dB.

Finally, the maximum allowed clock jitter is not a specification of the ADC itself, but a requirement for the clock that drives the converter. By jitter, we mean the cycle-to-cycle variation of the rising (falling) edge of the sampling signal. If the command used to trigger the sampling is affected by jitter, the sampling is not uniform. This uncertainty in the sampling instant is called "aperture jitter" and is usually measured in picoseconds rms, t_j. For a given value of the aperture jitter, the error introduced increases as the slope of the input signal, dV_{in}/dt, increases and causes a degradation in the SNR:

$$SNR = 20 \log\left(\frac{1}{2\pi f_{in} t_j}\right) \quad (4.1)$$

where f_{in} is the frequency of the input signal. For example, for the maintain the required accuracy in a 12-bit ADC for an input signal of 10 MHz, it is necessary that the rms jitter shall be 3.89 ps. The presence of jitter manifests itself as a uniform noise distributed over the band, while the clock frequency has no effect on the amount of jitter noise. The effect of jitter on SNR is accentuated in applications based on subsampling due to the high frequency of the input signal.

4.2.2 Main ADC architectures

Based on the mean used to convert the analog signal into digital, several ADC architectures, each addressing different resolution-speed requirements, have been proposed. A basic list of the most common ones is:

- Flash
- Pipelined
- SAR
- Ramp
- Sigma-delta

More than DAC, ADC can be classified into two main categories: Nyquist and oversampling converters. This is related to the ratio between the maximum analog input signal frequency and the clock frequency used for the conversion. As for definition, the sampling frequency of a Nyquist-rate data converter is normally near the Nyquist limit, that is, $f_{clk}/2$. The first five architectures belong to the category of Nyquist-rate, while the sigma-delta to the oversampling converters.

As for DACs, a complete description of all variants of these circuits is outside the scope of this book and can be easily found in existing literature [1], [3]-[4]. However, in the following, the basic principles behind their operation will be reviewed, also because the rad-hard versions of ADC start from traditional architectures.

Flash ADC

Flash ADCs operate using a parallel, distributed sampling approach that enables very fast conversion. Figure 4.7 illustrates the block diagram of an

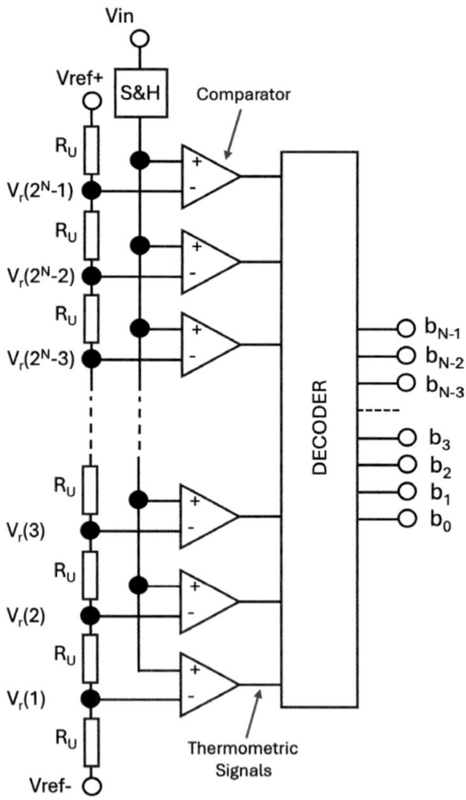

Figure 4.7 Flash ADC block diagram [1].

N-bit flash converter. The circuit is made up of a resistor ladder with 2^N equal segments, 2^{N-1} comparators, and a decoding stage. The resistor ladder provides 2^{N-1} evenly spaced reference voltages, while each comparator checks the input signal against one of these reference levels. The outputs of the comparators form a thermometer code (a string of logic "0"s followed by "1"s), which the decoder then translates into the corresponding binary output.

Thanks to their parallel operation, flash ADCs are conceptually the simplest and fastest converters. However, they also come with significant limitations. The main drawback is that the number of comparators increases exponentially with resolution – making 8 bits the practical upper limit. The same scaling applies to the resistors in the ladder network. Although the resistors in the divider are designed to be identical, both random and systematic mismatches occur in integrated resistors. As the resistor area grows, so does the impact of process gradients, which manifests as integral nonlinearity.

A related approach is the sub-ranging, or two-step, ADC. This architecture can be seen as a variation of the flash converter, designed to overcome its impracticality at resolutions beyond 8 bits, where area, power consumption, and input capacitance become excessive. Two-step converters trade some speed in exchange for improved area and power efficiency. They can be considered in the same way as the counterpart of the "intermeshed architecture" of DACs.

Alternative methods, such as folding and interpolation, enhance conventional flash designs. These approaches preserve the primary strength of flash converters – namely, single-cycle (or "one-step") conversion – while alleviating their large power and area requirements. Interpolation creates intermediate reference levels between known voltages, typically using resistive or capacitive dividers. This effectively halves the number of required preamplifiers and reference voltages, which in turn reduces both the input capacitive load and the size of the resistor ladder.

Folding extends the concepts of flash and two-step ADCs by adding analog preprocessing. It generates a residue signal from the input, which is digitized to extract the least significant bits (LSBs). At the same time, the most significant bits (MSBs) are obtained from a coarse flash stage operating in parallel with the folding circuit, both sampling the input simultaneously. Interpolation can also be incorporated into folding designs for further optimization.

Pipeline

Pipelining, a principle widely used in digital design, can also be applied to ADCs, where processing steps are carried out in sequence. In this context, pipelining means executing different parts of a sequential conversion

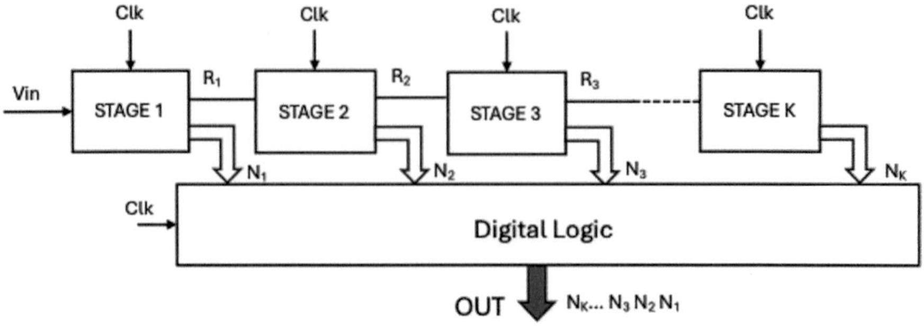

Figure 4.8 Block diagram of a *N*-stages pipeline ADC.

algorithm in parallel: several stages operate simultaneously, each performing one step of the algorithm.

A typical example of a sequential algorithm is the two-step ADC discussed earlier. In its basic form, this converter requires two clock cycles to determine first the most significant bits (MSBs) and then the least significant bits (LSBs). In the pipelined version, however, both parts are handled concurrently: while the coarse flash stage determines the MSBs of a new input sample, the residue amplifier processes the quantization error from the previous sample, allowing the fine flash stage to resolve its LSBs. This approach boosts throughput (and can also reduce area and power) but introduces a latency between analog input and final digital output.

A true pipeline ADC extends this concept further. Figure 4.8 illustrates the block diagram of a generic pipelined architecture. The first stage determines one or more MSB bits $\{N\}_1$; the next stages decide additional bit(s) $\{N\}_2$ and so forth until the last stage that estimates the LSBs $\{N\}_k$.

Each pipeline stage produces both a digital output and a residue, R_k, which serves as the analog input for the following stage. Typically, the first half of the clock cycle is devoted to sampling and bit decision, while the second half is used to generate the residue. The next stage must align its sampling instant with the moment the residue becomes available. As noted earlier, both the sampling and the final output occur at the clock frequency. This means that the bits generated by the second stage appear half a clock cycle later than those from the first stage, and the same timing relationship continues across subsequent stages. As a result, at any given clock instant, all stages are working on different samples in parallel. The digital correction logic then aligns these outputs using appropriate delay elements to assemble the final overall digital word.

Figure 4.9 shows the block diagram of a generic pipeline stage. The ADC generates j-bits. The DAC has the same number of bits.

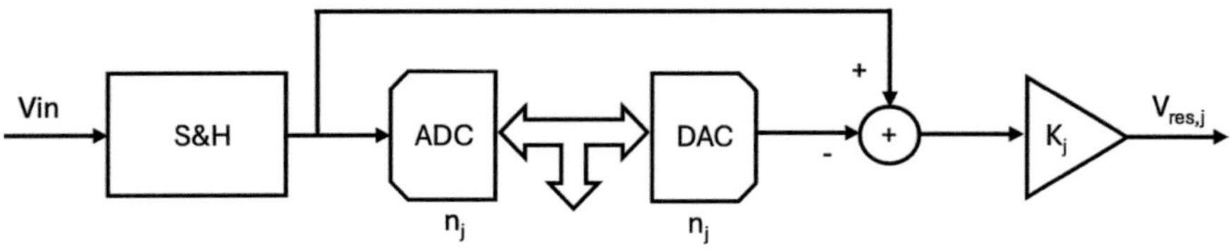

Figure 4.9 Block diagram of a pipeline stage.

The subtraction of the DAC output from V_{in} gives the quantization error of V_{in} quantized with the number of bits of the DAC. The amplification given by the inter-stage block increases the dynamic range of the residual. If the number of bits of the DAC is n_j and the gain is $K_j = 2^{nj}$, the dynamic range of the residual equals the dynamic range of the input. This condition permits the designer to use the same reference voltage in all pipeline states.

The offset, gain error, and nonlinearity of the input Sample and Hold stage directly translate into offset, gain error, and nonlinearity for the entire pipelined ADC. Likewise, imperfections in the internal ADCs, DACs, and residue amplifiers contribute to both differential nonlinearity (DNL) and integral nonlinearity (INL). Comparable limitations are also present in two-step converter architectures.

Successive approximation register (SAR)

The successive approximation ADC uses a binary search strategy implemented in a feedback loop that contains a 1-bit comparator. Its main purpose is not to maximize conversion speed but to keep the circuitry simple and to lower power consumption. This benefit comes at the cost of reduced speed, as will be discussed later. Figure 4.10 illustrates this architecture.

The Sample and Hold captures the input during the first clock cycle and maintains it for the following N cycles. The digital control logic then drives the DAC according to the successive approximation procedure. At the start of conversion, the most significant bit (MSB) is set to "1" as an initial guess. If the comparator validates this assumption, the bit is kept; otherwise, it is reset to zero. In the next clock cycle, the next bit (MSB–1) is tentatively set to "1." The comparator again checks the guess, and the process continues in the same way, step by step, until the least significant bit (LSB) is resolved. Meanwhile, the logic delivers the digital output while the Sample and Hold acquires the next input.

A key drawback is that if the comparator produces an incorrect decision at any stage, the error alters the search path and propagates through the remaining steps. This makes the architecture more vulnerable to single-event transients (SET).

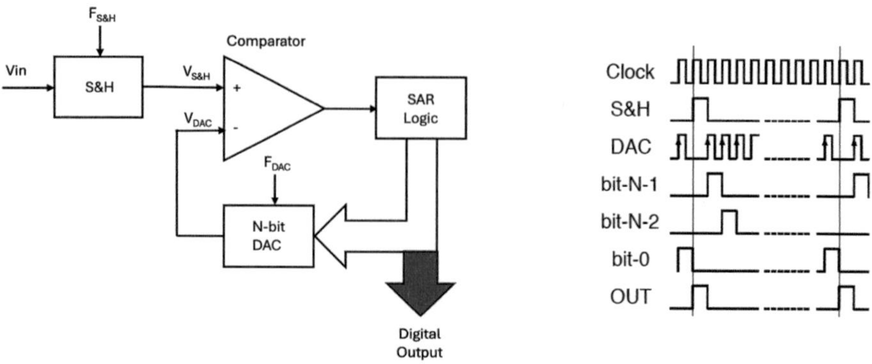

Figure 4.10 Basic architecture of a SAR converter and timing diagram.

For an N-bit resolution, a SAR ADC is slower than a full flash converter. This is because it requires one clock cycle for sampling plus one additional cycle for each of the N bits, leading to a total of $(N + 1)$ clock periods per conversion. On the positive side, the SAR approach does not need a subtractor (unlike a pipeline ADC), and both circuit complexity and power consumption are generally lower than in other architectures.

The accuracy of successive approximation converters depends on the accuracy of the Sample and Hold and DAC. Often, a charge redistribution technique is used in SAR that incorporates capacitors DAC [5].

Sigma-delta

The sigma-delta modulator is based on a specific algorithm to convert band-limited signals into a data stream. The idea is that a reduced accuracy per sample can be compensated for by using a high sample rate to convert a limited bandwidth. Sigma-delta converters offer high-resolution, high integration, and low cost, making them a good architectural choice for applications such as process control and monitoring. A sigma-delta ADC converts an analog signal into a digital signal by oversampling and using a feedback loop (Figure 4.11).

The analog portion of a sigma-delta converter (in principle, a 1-bit ADC) is very simple. The digital side, performing filtering and decimation, makes the sigma-delta ADC inexpensive to produce, but it is more complex. The oversampling method becomes more effective if the noise spectrum is lowered in the signal band, possibly at the expense of an increase in the out-of-band portion, thereby changing the white spectrum of the quantization noise into a shaped spectrum. Having more noise in high-frequency regions is not problematic, as the digital filter used after the ADC removes it (Figure 4.12).

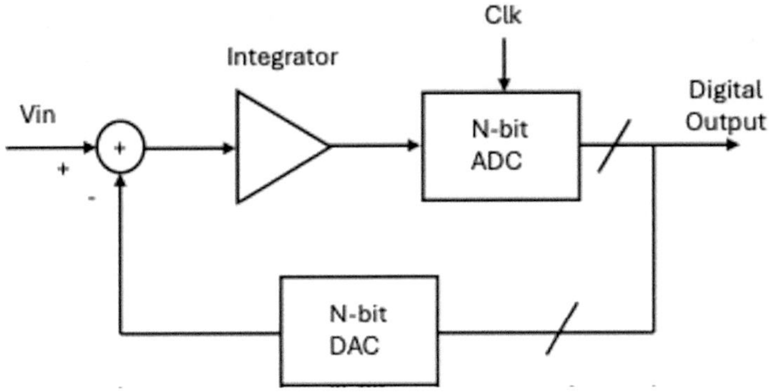

Figure 4.11 Block diagram of a sigma-delta modulator.

Incorporating the quantizer in a feedback loop, as shown in the figure, gives rise to the desired in-band noise reduction, also called noise shaping.

The sigma-delta works according to the following steps:

- Oversampling: The input is sampled much faster than needed.
- Integration: The signal is accumulated over time.
- Delta comparison: A comparator (in principle a 1-bit ADC) outputs 1 or 0 depending on the signal level.
- Noise shaping: The feedback loop pushes noise into higher frequencies.
- Digital filtering and decimation: A digital filter cleans the signal and reduces the sample rate.

First, consider the frequency-domain transfer function of a traditional multi-bit ADC with a sine-wave input signal. This input is sampled at a frequency of F_s. According to Nyquist theory, F_s must be at least twice the bandwidth

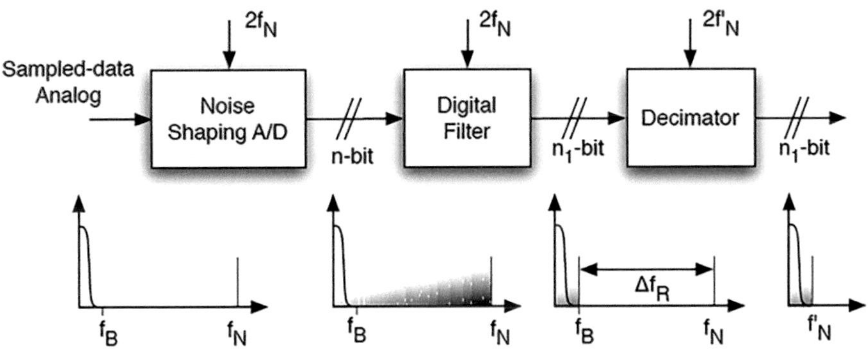

Figure 4.12 Out-of-band noise rejection and decimation of an oversampled signal [1].

of the input signal, f_B. When observing the result of a fast Fourier transform (FFT) analysis on the digital output, we see a single tone and lots of random noise extending from DC to $F_s/2$. Known as quantization noise, this effect results from the following consideration: the ADC input is a continuous signal with an infinite number of possible states, but the digital output is a discrete function whose number of different states is determined by the converter's resolution. So, the conversion from analog to digital loses some information and introduces some distortion into the signal. The magnitude of this error is random, with values up to ±LSB. If we divide the fundamental amplitude by the root mean square (RMS) sum of all the frequencies representing noise, we obtain the signal-to-noise ratio (SNR). For an *N*-bit ADC, SNR = 6.02 N + 1.76 dB. To improve the SNR in a conventional ADC (and consequently the accuracy of signal reproduction), you must increase the number of bits. In a Sigma-Delta ADC, quantization noise is not simply reduced but spectrally shaped. Through the use of a feedback loop with one or more integrators, the modulator forces most of the quantization noise toward higher frequencies, outside the signal band. As a result, the noise power within the signal bandwidth is significantly reduced. This process is known as noise shaping. In addition, the converter operates with oversampling, spreading the total quantization noise over a wider frequency range. The combination of oversampling and noise shaping greatly improves the in-band signal-to-noise ratio. Finally, the digital decimation filter removes the out-of-band noise, leaving a higher-resolution digital output.

Ramp converters

Single-slope and dual-slope converters are two classic ramp converters, especially used in applications where simplicity and precision are more important than speed.

The single-slope ADC works by comparing the input voltage to a linearly increasing reference voltage, often generated using a constant current charging a capacitor (i.e., a ramp, see Figure 4.13(a)). Their operation is described by the following steps:

- Start a timer: A digital counter is started at the same time a ramp generator (done with an integrator) begins to produce a steadily increasing voltage V_C.
- Compare voltages: A comparator continuously checks when the ramp voltage reaches the input analog voltage.
- Stop the timer: When the ramp voltage equals the input voltage, the comparator sends a signal to stop the counter.

- Digital output: The final count value represents the input voltage in digital form. A higher input voltage takes longer to reach, resulting in a higher count.

The dual-slope converter charges a capacitor with the input voltage for a fixed amount of time, then discharges it using a known reference voltage. The time it takes to return to zero is measured and used to compute the input voltage (see Figure 4.13(b)). Their functionality is articulated as follows:

- Integration phase (charging):
  - The ADC connects the input analog voltage to an integrator for a fixed time T1
  - The integrator output ramps up (or down) depending on the input voltage level.

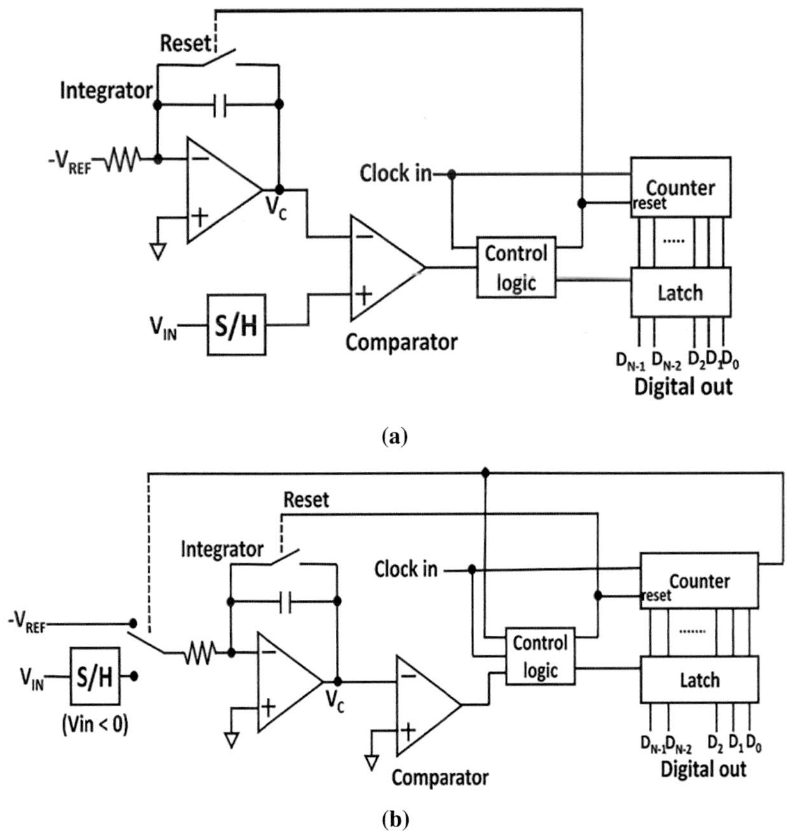

Figure 4.13 (a) Single-slope converter [1]; (b) dual-slope converter.

- De-integration phase (discharging):
  - The input is disconnected and replaced with a known reference voltage of opposite polarity.
  - The integrator output begins to ramp back toward zero.
  - A timer counts how long it takes (T2) to return to zero.
- Digital output:
  - The time T2 is directly proportional to the input voltage.
  - The ADC outputs a digital value based on this measured time.

4.3 Rad-hard Analog-to-Digital Converters

The radiation-hardened analog-to-digital converter serves as a bridge between the external analog environment and digital circuits. Its performance typically ranges from a few kilohertz up to several hundred megahertz in terms of sampling rate, and from a few bits up to 16 (or more) bits in resolution. To meet these specifications, various architectural approaches are considered, each tailored to satisfy specific electrical demands.

Moreover, to operate in a harsh environment, the ADC must be designed to withstand radiation effects, which imposes constraints that guide the choice toward certain architectural solutions. As discussed earlier, the literature presents several ADC topologies, each optimized for different resolution and speed requirements. For reach of them, conventional ADC architectures – such as flash, SAR, ramp, sigma-delta, and pipelined – serve as the foundation for the design of a rad-hard counterpart.

On this matter, a lot of literature is available nowadays ([6]-[18] and [24]-[52]). Figure 4.14 represents the percentage of different topologies of rad-hard ADCs calculated over the most significant and recent 62 papers.

As can be seen, the most popular architectures are SAR and pipelined, while in third place, there is the flash topology. These topologies are well-suited to convert fast-varying signals, while sigma-delta and ramp topologies are mainly used to convert slow-varying signals, such as audio or image sensors. As CMOS technology continues to scale down, there has been a shift in design philosophy: analog-to-digital signal conversion is increasingly being performed in the time domain. This trend is driven by the fact that, as transistor sizes shrink, supply voltages also decrease, which in turn limits the

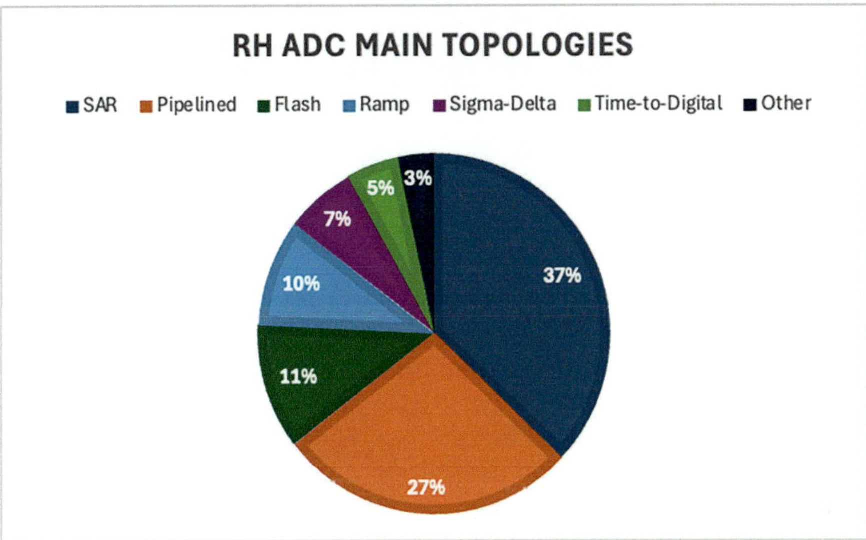

Figure 4.14 Percentage of the most popular RH ADC architectures calculated over 62 papers.

dynamic range of traditional ADCs. However, modern nanometer-scale processes offer excellent high-frequency characteristics, making it possible to achieve greater precision in the time domain. In this domain, analog information is represented by the time interval between signal zero-crossings. Instead of a conventional ADC, a time-to-digital converter (TDC) is used. TDC finds application in a wide range of areas such as all-digital phase-locked loops (ADPLLs), communication systems, localization and positioning technologies, sensor interfaces in measurement equipment, time-of-flight systems, digital oscilloscopes, particle physics experiments, and beyond. This book covers analog-to-digital converters where the input is an analog voltage only. So that TDC is not included, while a good overview can be found in Refs. [6], [7].

From the above discussion, it emerges that each different ADC architecture has the characteristics that make it suitable for different applications. However, they shall be assessed with respect to radiation since they exhibit different behavior as a response, as well. Table 4.1 summarizes the radiation sensitivity (TID tolerance and SEE tolerance) as a function of ADC structures.

Table 4.1 Sensitivity comparison of various ADC topologies against TID and SEE.

ADC topology	SAR	Pipelined	Flash	Ramp	Sigma-delta
TID tolerance	😐	☹	😐	😐	☺
SEE tolerance	😐	☺	☺	😐	😐
RHBD provisions	⇔	⇧	⇩	⇔	⇔

The meaning of the last row is that there are design levels where the intervention to make the ADC resilient shall be higher (arrow up). In general, the greater the complexity, the higher the sensitivity to radiation, especially to dose.

Among them, pipelined designs incorporate a significant amount of analog circuitry, making them particularly vulnerable to radiation-induced variations. Components like amplifiers and comparators, as noted in [8], exhibit issues such as offset, gain instability, and inaccuracies in current mirrors caused by total ionizing dose. To address these challenges, specialized design techniques have been developed [9]. On the other hand, since an internal digital correction logic aligns and combines the bits from all stages to produce the final digital word, intrinsically this redundant logic for fixing of errors can also be extended to overcome SEU.

SAR converters also experience offset problems. When implemented using charge redistribution methods, TID introduces an additional effect: the threshold voltage of NMOS transistors decreases, leading to increased leakage currents at the top plate of the capacitor array [10]. Moreover, SAR ADCs are especially susceptible to single-event upsets. If an SEU disrupts the control logic during conversion, it can result in a bit flip. Since the switching sequence relies on previously determined bit values, a single error can cascade, causing multiple bits to be incorrect. A notable approach that aims to mitigate these issues involves an ADC architecture using two DACs; however, this solution significantly increases both area and power consumption [11]. Lastly, sigma-delta ADCs represent a promising solution for radiation-hardened applications. The analog section of the circuit is expected to be less affected by radiation-induced disturbances, since, in principle, any faults caused by excess charge in the integrator capacitors are shaped by the modulator into out-of-band noise [7] [12].

The single-slope (or dual-slope) architecture is preferred for detectors in modern physics experiments and in image sensors thanks to its simplicity, its low power consumption, and small overall area [13]-[16]. It is also perfectly adapted to multichannel applications, where conversion speed is not important. In this architecture, two sensitive points can be identified with respect to SEE: SEU in sequential logic (counters and latches) and SET in the ramp generator (generally an integrator implemented with an integrated capacitor). For the logic, the provisions are the same as those for the digital section in any converter. The unwanted charge injection caused by SEE in the capacitor can make an incorrect one-sample conversion.

Finally, the flash topology demonstrates good robustness against radiation effects. This approach was widely adopted until the late 1990s due to its simplicity and modular design [17]. Indeed, from a radiation-hardening perspective, this architecture offers several inherent benefits:

- It follows a largely digital-oriented design methodology.
- It supports parallel signal processing.
- Each comparator integrates its own distributed sample-and-hold functionality.

The flash ADC primarily consists of digital components, including the comparators, which played a key role in choosing this topology. To these, digital cells can be made radiation-tolerant by applying RHBD techniques [18]. Furthermore, the flash structure allows parallel processing of the input signal. This means that a fault in one comparator stage does not affect the entire output word and can be corrected using either standard correction techniques or the more advanced method proposed in this work. This represents a notable advantage compared to, for example, SAR ADCs. Finally, the list of reference voltages is usually implemented by means of resistors. Fortunately, resistors in polysilicon are essentially insensitive to radiation, as reported about DAC in Chapter 3.

Several tests have been done in the past on ADC commercial components, showing to what extent TID and SEE can affect device performance [19]-[23]. It is worth noting that the main part of commercial components is radiation tolerant or radiation hard only up to some tens of krad (Si). An example of the level of resiliency reached by some commercial components in terms of TID and SEL is reported in Table 4.2. It is worth noting that it does not pretend to be exhaustive.

Several significant examples of ADCs described in published papers will be shown below, corresponding to the most popular architectures. Finally, the last paragraph will describe a SAR ADC proposed and developed by the authors.

Table 4.2 Main commercial rad-hard ADC components

Part	Manufacturer	Speed resolution	Topology	TID	SEL LETth (MeVcm^2/mg)
ADC14155QML	Texas Instruments (CMOS)	155 MSPS 14-bit	Pipelined	100 krad (Si)	120
ADC10D1000QML	Texas Instruments (CMOS)	1–2 GSPS 10-bit	Folding-interpolating	100 krad (Si)	120
ADC12D1620QML	Texas Instruments (CMOS)	1.6–3.2 GSPS 12-bit	Pipelined	300 krad (Si)	QML-V qualified
ADC128S102QML	Texas Instruments (CMOS)	50 kSPS–1 MSPS 12-bit	SAR, multichannel	Space-grade optimized	Space-grade adapted
ADC168M102R	Texas Instruments	1 MSPS 16-bit	SAR	Radiation-tolerant	Radiation-tolerant
ISL73148SEH	Intersil-Microrel (0.25 um CMOS)	~1 MSPS 14-bit	SAR	75 krad (Si)	>86
RHF1201/1401	STM (CMOS)	20–50 MSPS 12–14-bit	Pipeline	300 krad (Si)	120
RHFAD128	STM (CMOS)	50–1000 kSPS	SAR	300 krad (Si)	125
Vaious components	Frontgrade Technologies	Various speed	Various	Up to 1 Mrad (Si)	

Flash rad-hard ADC

A first proposal of rad-hard ADC is in [24]. It presents the design and implementation of an 8-bit, 20 MHz flash analog-to-digital converter (ADC) with radiation hardness up to 100 MRad (SiO_2, 10 KeV X-ray). It leverages the SIMOX SOI CMOS technology to combine high-speed analog performance with robustness against ionizing radiation. The paper also analyzes the trade-offs between radiation hardness and electrical performance compared with non-hardened versions of similar converters.

The ADC was built using SOI technology, with a 1.2 μm active silicon thickness and 0.3 μm buried oxide. The ADC has been designed following RHBD techniques. The general architecture is not reported here because it is quite similar to that in Figure 4.7. Circuit architecture included 256 comparators, a dual auto-zero technique, and master-slave flip-flops to mitigate threshold shifts due to irradiation. The comparators are auto-zeroed type (Figure 4.15), so that the auto-zeroing compensates for threshold variations. Indeed, it maintains operation at the inverter's high-gain region, automatically compensating for dynamic threshold variations.

Another adopted provision is the use of 1.4 μm channel lengths in comparator MOS transistors to improve hardness. Moreover, a static ROM was used for thermometer-to-binary conversion instead of a pre-charge PLA (Programmable Logic Array), reducing sensitivity to threshold variations. Finally, output buffers were designed with ELT and body connections to suppress leakage after irradiation.

Electrical performance was assessed before and after exposure to X-ray irradiation (up to 100 Mrad at 400 krad/min). Measurements were taken within 10 minutes after irradiation, with the circuit powered and stimulated (1 MHz analog input, 20 MHz clock). The hardened ADC achieved reliable operation up to 20 MHz under irradiation, as shown in Table 4.3.

Test results show a higher dependence of INL and ENOB on total dose for high frequency, due to threshold voltage shifting of MOS (edgeless transistors only in output buffers).

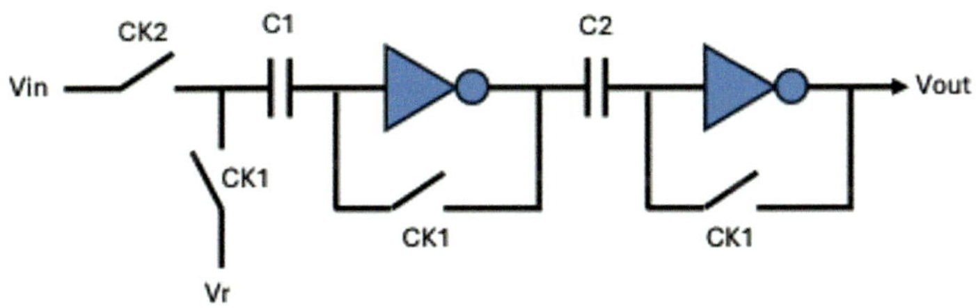

Figure 4.15 Auto-zeroed comparator used in [24].

Table 4.3 Comparison between conventional and rad-hard versions of the ADC [24].

Parameter	Conventional	Rad-hard
Max input frequency	10 MHz	10 MHz
Max clock frequency	25 MHz	20 MHz
INL (pp)	1.2 LSB	1.6 LSB
ENOB	7.5	7.4
SNR	44 dB	42.5 dB
Power consumption	250 mW	450 mW

Compared to a non-hardened version main penalties are:

- Chip area increased by ~50%.
- Power consumption nearly doubled (250 mW → 450 mW).
- Maximum speed decreased by ~20%.

At low frequencies (≤3 MHz), ADC performance was almost unchanged after 100 Mrad.

A second example of a flash converter in Ref. [25] is quite similar in architecture to [24]. Its main goal was to combine radiation-hardened CMOS/SOS (Silicon-On-Sapphire) processing with a novel comparator-inverter design to achieve 6-bit flash ADCs capable of withstanding up to 1 Mrad (Si) total-dose radiation. The ADC architecture was based on parallel comparators (64 for a 6-bit design), each comparing the input signal against reference voltages generated by a polysilicon resistor ladder. Comparator outputs generated a thermometer code, which was decoded into binary form and latched into output buffers.

Instead of conventional CMOS comparators, an auto-zeroed comparator similar to [24] was used. The autozero comparator was also tolerant of leakage increases, since leakage currents remained negligible compared to bias currents.

Radiation hardness was enhanced by fabricating circuits in a rad-hard CMOS/SOS process:

- Devices built on sapphire substrates (isolation against latch-up, SEUs, transients).
- Special channel-oxide process for megarad hardness.
- Self-aligned polysilicon gates to minimize capacitances.

The test methodology included two flash ADCs fabricated both in standard CMOS/SOS and rad-hard CMOS/SOS processes, respectively. The measured performance before radiation is reported in Table 4.4.

Devices were exposed to Co-60 gamma irradiation up to 1 Mrad (Si). Electrical measurements included leakage currents, offset and gain errors, linearity, and functional checks. Rad-hard devices remained fully functional up to 1 Mrad, while standard devices failed at doses as low as 5–45 krad. In rad-hard version:

- Supply current increase was minimal (≈2.5 mA at 5 V, 1 Mrad), compared with large increases in standard devices.
- Tri-state leakage currents increase to only ~1 μA at 1 Mrad, versus ~1 mA at 45 krad in standard versions.
- Functional checks confirmed continuous ADC operation up to 1 Mrad, with all parametric limits met at +25 °C.
- Offset and gain errors stayed within ±1 LSB across the full dose range.
- Differential and integral linearity remained within ±0.5 LSB up to 1 Mrad.

Some devices showed slight increases in data delay at high doses, which could affect maximum operating frequency, but still within acceptable limits.

Standard process devices failed due to NMOS channel leakage and reduced thresholds, preventing proper comparator operation. The rad-hard process significantly reduced threshold voltage shifts (<0.25 V per 100 krad) and suppressed leakage increases. Overall, the autozero comparator design and rad-hard process proved complementary, providing resilience against both circuit-level and device-level radiation effects.

A third example of a flash converter belonging to the same class of architectures as the first two is in Ref. [26]. It is an 8-bit full-flash whose resiliency to radiation is mainly obtained through process: 2 um SOS and

Table 4.4 Measured performance of flash ADC in [25].

Parameter	Value
Value	6 bit
Resolution	15 MHz
6 bit	0.35 LSB
Max clock frequency	0.15 LSB

auto-zeroing comparator, too. Performance degrades significantly after 50 krad 60Co source Gamma radiation in terms of digital leakage output current, analog current consumption, and INL since any ELT has been used.

To improve the rad-hard performance, a variation of the typical flash converter architecture can be found in Ref. [27]. The study investigates how to improve the single event upset tolerance of flash ADCs in radiation-heavy environments.

Indeed, standard flash ADCs are vulnerable to SEU effects in both analog and digital sections, leading to output errors. Here, the goal was to propose structural design changes (in line with circuital RHBD methodology) that mitigate SEU without degrading ADC performance by:

- Minimize bubble errors in thermometer code caused by comparator upsets (A-SEU).
- Prevent SEU propagation in clock distribution and logic blocks (L-SEU).
- Preserve accuracy, speed, and power efficiency in current CMOS technologies.

These targets are obtained first by analyzing the conventional flash ADC architecture and restructuring it, focusing on three critical blocks:

- Comparator stage (analog-sensitive).
- Binary tree clock distribution (digital-sensitive).
- Flip-flop latches and output logic.

For A-SEU (SEU located in the analog comparator) hardening, a corrective logic was added at the thermometer-to-binary encoding stage: instead of simple $(\overline{C_n} * C_{n+1})$ operations, a modified formula $(C_{n-1} * \overline{C_{n+2}} * (C_n \text{ xor } C_{n+1}))$ corrected bubble errors (where C_n and C_{n+1} are respectively the *n* and *n*+1 comparators output logic levels). This ensured valid "1-of-N" codes even under multiple comparator upsets (Figure 4.16).

For L-SEU (SEU located in latches), hardening of the binary tree:

- A new dissociated binary tree (DBT) structure was introduced (Figure 4.17).
- DBT isolates upset propagation, so errors remain separated and correctable by the logic.

The used inverter was reinforced by large transistor sizing and RC filtering.

When combined with the previously described corrective logic, the DBT ensures that if an upset occurs at any node of the binary tree inverter,

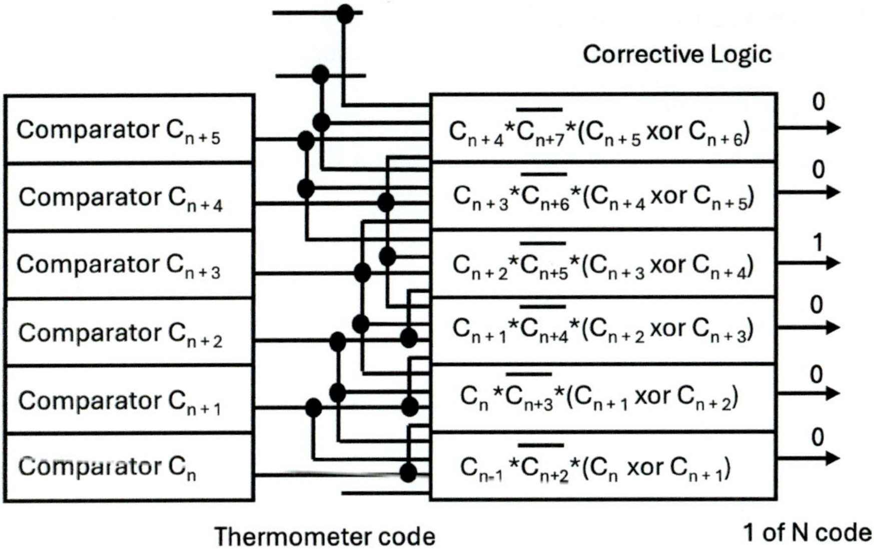

Figure 4.16 Corrective logic removing bubble errors [27].

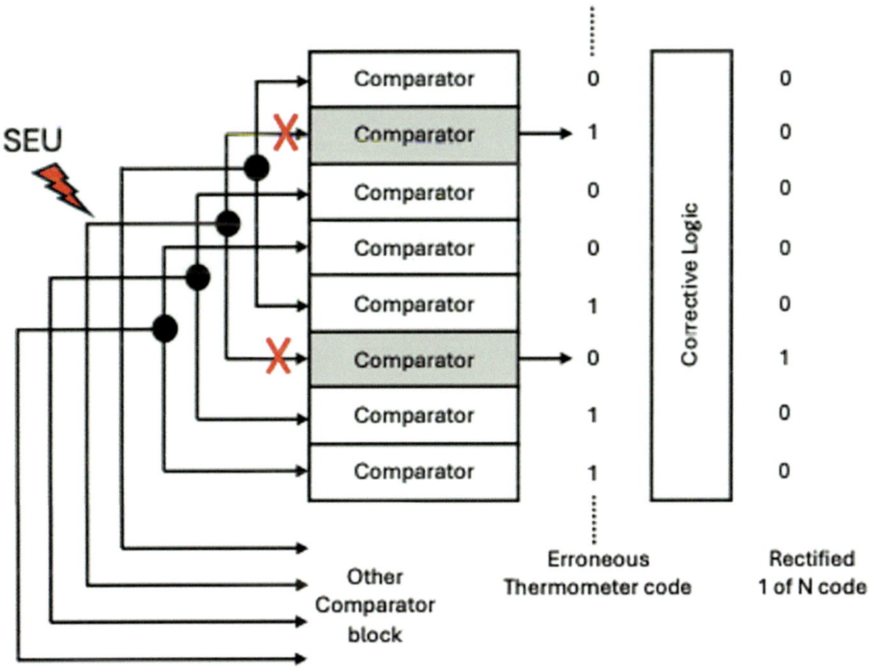

Figure 4.17 Correction of false comparisons due to an SEU rising on a binary tree inverter [27].

the resulting error is isolated by more than three comparators, allowing the corrective logic to restore the correct 1-of-N code.

For Latch hardening:

- Hardened flip-flops were designed with RC filtering and high-impedance nodes.
- Proven improved linear energy transfer (LET) thresholds (30 MeV*cm^2/mg vs. 6 for standard).

For ROM encoder hardening:

- Pre-charge buses were eliminated to avoid high-impedance states.
- "1" levels generated by PMOS and "0" by NMOS, ensuring robustness against leakage and threshold shifts.
- High bus capacitance naturally suppressed transient SEUs.

Comparators were further protected by cyclic auto-zero analog structures for TID hardening.

The converter has been implemented in 0.6 μm CMOS technology, and post-layout simulations showed that corrective logic effectively eliminated bubble errors, limiting SEU impact to a maximum of 1 LSB error. The DBT reduced SEU propagation in clock trees, enabling robust correction by subsequent logic.

Hardened latches showed significant SEU resistance:

- LET threshold , 30 MeV*cm^2/mg for hardened vs. 6 MeV*cm^2/mg for standard.
- Cross-section reduced by more than one order of magnitude.

Encoder hardening prevented functional failures, maintaining reliable binary outputs despite transistor degradation. Simulation of SEUs confirmed all proposed techniques were effective at the block and system level. Two identical flash ADCs, hardened and un-hardened, have been designed. The performance comparison between hardened and unhardened ADCs is in Table 4.5.

The effective number of bits remained comparable (~5.6–5.9 bits). Signal-to-noise and distortion stayed within 35–37 dB, slightly better in hardened versions. Harmonic distortion was not significantly worsened.

Table 4.5 Comparison performance between hardened and un-hardened ADC [27].

Sampling frequency (MHz)	ADC converter	SINAD dB	Harmonic distortion dB	Effective bits number
100	Un-hardened	35.8	–39.87	*5.65*
100	Hardened	36.5	–42.6	*5.77*
50	Un-hardened	37.45	–46.75	*5.93*
50	Hardened	36.85	–42.6	*5.83*

Hardened ADCs preserved silicon area and power efficiency, avoiding the penalties usually associated with redundancy.

To assess the effectiveness of the hardening concepts, a current pulse carrying up to 5.5 pC of charge (with a typical amplitude of 10 mA, 500 ps duration, and 50 ns rise and fall times) has been applied to sequentially emulate upsets across the various blocks of both hardened and unhardened designs. All proposed hardening techniques have been validated by the paper's authors, considering the actual model card parameters of the commercial technology used.

Another variation on flash topology can be found in Ref. [28]. The paper proposes a redundancy-based, fault-tolerant methodology for analog circuits, focusing on a flash ADC architecture. The central idea is to use redundant comparators with an innovative analog voter to achieve high reliability without sacrificing performance. To achieve that, the authors of the paper aim to evaluate different voting strategies, quantify reliability improvements, and explore trade-offs between reliability, cost, area, and power.

A 4-bit flash ADC with 16 comparators was used as the baseline. Redundancy was introduced by replacing each comparator with an N-modular redundancy (NMR) block:

- Multiple redundant comparators.
- An analog voter to decide the correct output.

Two voting mechanisms were implemented:

- Majority voting: identifies the most common amplitude among inputs.
- Median voting: selects the median amplitude, which scales more efficiently at high redundancy levels.

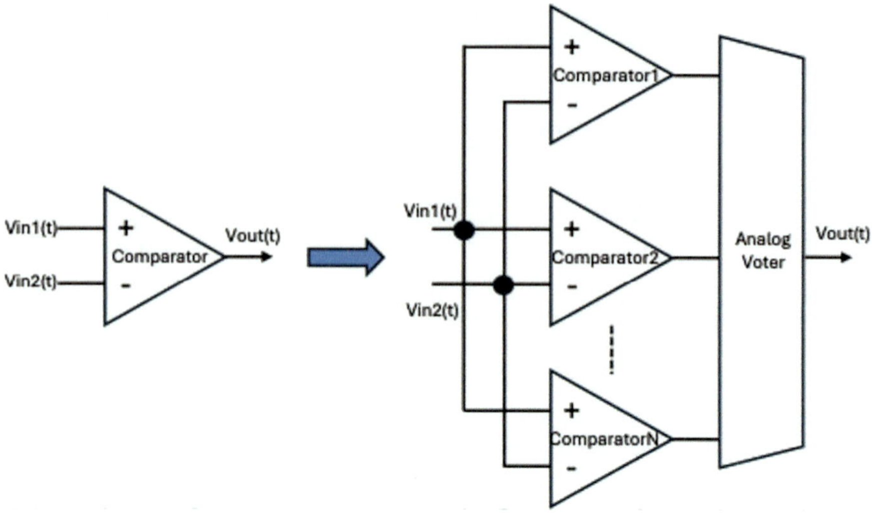

Figure 4.18 NMR comparator [28].

To design a fault-tolerant ADC, we replace all comparators with the NMR version of them, i.e., N comparators and a voter (Figure 4.18).

In this way, we can increase the reliability of this comparator by increasing the redundancy factor N. HSpice® simulations were performed using IBM 130 and 180 nm libraries to measure area, power, delay, and reliability. The paper demonstrates that analog voting with redundancy is a viable and effective method for fault-tolerant ADC design. Both majority and median voting strategies improve reliability, with median voting better suited for scalability.

Finally, the paper in Ref. [29] investigates how total ionizing dose irradiation affects high-voltage ADCs fabricated with thick-oxide MOSFETs. After identifying the mechanisms of threshold voltage (V_{TH}) shift and leakage current increase that degrade ADC accuracy, the authors describe the development of a radiation-tolerant ADC based on commercial CMOS SOI processes, avoiding costly process modifications. They propose a circuit- and layout-level reinforcement strategy exploiting the body effect in multi-voltage domains. Their target is to improve ADC TID tolerance from 20 krad (Si) to at least 100 krad (Si) while maintaining INL/DNL performance.

To achieve that, they started from a conventional 8-bit ADC composed of two 4-bit flash ADCs, operating in unipolar (0–5 V) or bipolar (±2.5 V) modes (Figure 4.19(a)). Irradiation test of unreinforced ADCs with Co-60 γ-rays, dose rate 50 rad (Si)/s, total dose up to 30 krad. The observed functional failures were linked to n-MOSFET sampling switches, whose V_{TH} becomes negative after TID.

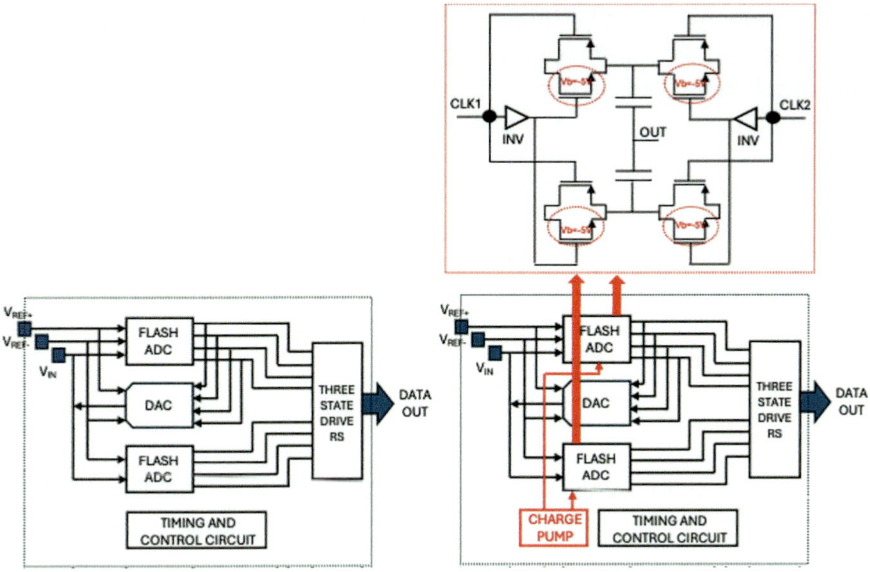

(a) (b)

Figure 4.19 Block diagram of the: conventional flash ADC (a); reinforced flash ADC (b) [29].

The proposed reinforcements (Figure 4.19(b)) are:

- Charge pump unit generating –5 V substrate bias in unipolar mode, exploiting the body effect to counteract TID-induced V_{TH} reduction.
- Enclosed gate layout to suppress edge leakage paths.
- Isolation ring to block parasitic MOSFET leakage.

Simulation studies in Cadence® confirmed minimal impact of the charge pump on timing and power. The reinforced ADCs have been fabricated in a 0.35 μm commercial CMOS SOI process. The experimental validation and comparison have been done with an irradiation up to 100 krad (Si), and the test of INL, DNL, ENOB, and power across multiple supply modes and temperatures.

The unreinforced flash ADCs:

- INL drift reached −18.11 LSB after 30 krad.
- DNL showed similar degradation.
- Failures traced to abnormal sampling when the V_{TH} of nMOS < V_{GS}.

Firstly, Authors tested the influence of total dose on an individual transistors. However, the reinforced devices, i.e., ring-gate nMOS, reduced ΔVTH after irradiation compared to strip-gate, but are insufficient alone. Then, they tested the parameters of the reinforced ADCs, and the TID experiments showed that:

- INL drifts only 0.51 LSB after 30 krad, 0.94 LSB after 80 krad, and 1.68 LSB after 100 krad.
- Improvement of more than 10 times over unreinforced devices.
- DNL variations remained within acceptable limits.
- ENOB and power consumption are essentially unchanged relative to baseline.

Moreover, the charge pump provides a stable –5 V bias without enlarging the layout area. Switch opening time only marginally increased; normal ADC speed preserved.

All in all, the reinforced ADCs show robust performance across both unipolar and bipolar domains. These ADCs achieved 100 krad (Si) tolerance, a fivefold increase over unreinforced (20 krad), and the solution is fully compatible with commercial CMOS SOI processes, requiring no special fabrication.

Sigma-delta rad-hard ADC

A first proposal of rad-hard sigma-delta ADC can be found in Ref. [12]. The authors propose the design of a very high-resolution ADC tailored for space applications. The main goal is to address the limitations of currently available space-qualified ADCs, which rarely exceed 16-bit resolution and often suffer from high power consumption. The proposed device aims to achieve effective resolution well beyond 16 bits while maintaining low noise and low power operation. It is designed to cover an exceptionally wide frequency range, from 0.1 mHz up to 1 kHz, enabling precise measurements of slowly varying physical parameters, as well as reliable spacecraft housekeeping tasks. Additionally, the design targets immunity to TID degradation and SEE, making it suitable for long-term use in space missions.

The proposed ADC is built on a second-order discrete-time sigma-delta (ΣΔ) modulator with a 1-bit quantizer (Figure 4.20). The oversampling ratio is selectable between 64 and 2048, allowing flexible trade-offs between speed and accuracy. The nominal output sampling frequency is 6 kHz, extendable up to 96 kHz. To mitigate flicker (1/f) noise and offsets, correlated

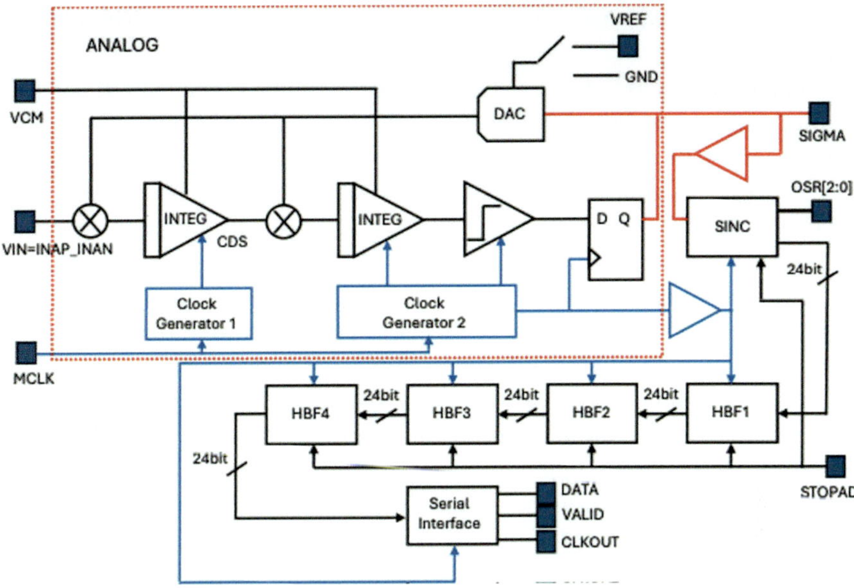

Figure 4.20 Sigma-delta ADC block diagram [12].

double sampling (CDS) is employed, effectively performing auto-zeroing. The design is realized using a 0.15 μm CMOS-on-SOI radiation-hardened technology.

The ΣΔ modulator itself relies on switched-capacitor (SC) integrators, chosen for their robustness against process variation and clock jitter. Operational transconductance amplifiers (OTAs) are optimized for operation at 6.144 MHz, balancing gain-bandwidth with low-noise requirements. The quantizer is implemented as a regenerative latched comparator offering microvolt-level sensitivity. Downstream, a decimation filter processes the modulator output, consisting of a fourth-order SINC filter followed by four cascaded half-band filters. Together, these filters guarantee suppression of aliasing bands down to −140 dB. For radiation hardening, the analog part of the circuit is intrinsically immune to radiation effects, because ideally all faults introduced by extra-charges in integrator capacitors are transformed into out-of-band noise by the modulator. Moreover, triple modular redundancy (TMR) and a finite state machine, along with a voting scheme is used in the digital section. The analog circuits employ ELT and guard rings. Deep trench isolation (each CMOS structure is isolated using a deep trench extending down to the buried oxide of the SOI) further ensures immunity to latch-up, with design specifications exceeding 70 MeV mg/cm² for SEL and 100 krad (Si) for TID.

The design was validated through extensive post-layout simulations (Table 4.6):

The ADC demonstrated a spurious-free dynamic range (SFDR) of 110 dB. Simulations confirmed an effective resolution of 18 bits, with theoretical analysis suggesting the potential to exceed 22 bits under ideal conditions. The achievable signal-to-noise ratio was limited by the chosen sampling capacitance of 0.7 pF, compared to the larger value required for maximum theoretical SNR. Nonetheless, the effective bandwidth extends from DC up to 16 kHz, comfortably surpassing the rated 1 kHz specification. The device supports sampling rates from 6 to 96 kS/s. Power consumption is modest, with the modulator drawing 2.2 mA from a 3.3 V supply, complemented by a 1.8 V supply for the digital logic. The overall chip, including I/O pads, occupies an area of 9.0 mm^2. Output data is provided as 24-bit serial words, ensuring high fidelity in digital interfacing. Simulations confirmed the effectiveness of the noise-shaping behavior of the ΣΔ modulator and demonstrated the benefits of CDS in reducing low-frequency noise, which is crucial for sub-Hz operation. Unfortunately, no experimental characterization is available up-to-now to validate the proposed provisions against radiation.

A second example of rad-hard sigma-delta can be found in [30]. It is focused on single-event upsets caused by charged particles, which are a significant challenge. Traditional high-resolution ADCs, although effective in conventional applications like audio, video, and instrumentation, are vulnerable to SEUs, particularly in their digital circuitry. The goal of this solution is twofold: first, to demonstrate that the analog ΣΔ modulator is inherently immune to SEUs in the modulator because radiation-induced faults manifest as high-frequency noise that can be filtered out; and second, to investigate strategies to protect the digital filtering stage, which is more vulnerable to

Table 4.6 Performance summary of sigma-delta ADC in [12].

Parameter	Value
Sampling rate	6 to 96 kS/s
Bandwidth	Up to 1 kHz
Clock frequency	6.144 MHz
SFDR	110 dB (18bit)
Power consumption (modulator only)	2.2 mA
Estimated TID	100 krad
Estimated LET for SEL	70 MeV

SEUs. By doing so, the authors intend to propose an ADC architecture that combines high performance with resilience to radiation effects, ensuring its viability for space applications.

To achieve these objectives, the authors developed a complete Matlab® model of a ΣΔ ADC (Figure 4.21). The model represents the modulator as a discrete-time accumulator and comparator, transforming input samples into binary sequences. The analog part of the converter, largely consisting of the ΣΔ modulator, was studied under simulated radiation conditions by injecting impulsive charges into the integrator capacitor, reproducing the effect of particle hits. For the digital section, which includes the decimation and low-pass filters, different implementations were tested under simulated bit-flip faults to evaluate their resilience to SEUs. The considered architectures included finite impulse response (FIR) filters, infinite impulse response (IIR Type 1) filters with the shift operator, and IIR (Type II) filters using the delta operator. The comparison focused on the trade-off between robustness to radiation, implementation complexity, and required silicon area. The delta operator approach was of particular interest, as it allows designing filters with equivalent frequency response but with fewer registers and reduced word lengths, thus decreasing the vulnerable area.

The simulation results confirmed that the analog ΣΔ modulator is naturally immune to SEUs. Radiation-induced current pulses in the modulator were translated into additional high-frequency noise, which is already characteristic of ΣΔ operation and therefore removed by the subsequent digital filtering stage. As a result, the modulator does not require additional protection measures. Conversely, the digital part proved to be highly sensitive to SEUs. In the decimation stage, bit-flips in critical registers introduced significant in-band errors, corrupting the signal. Among the low-pass filter options, the FIR filter showed moderate robustness to radiation effects but required a large silicon area due to its high number of taps and registers. The IIR filter with the conventional shift operator was highly sensitive, with faults severely degrading the signal, often making the fundamental frequency

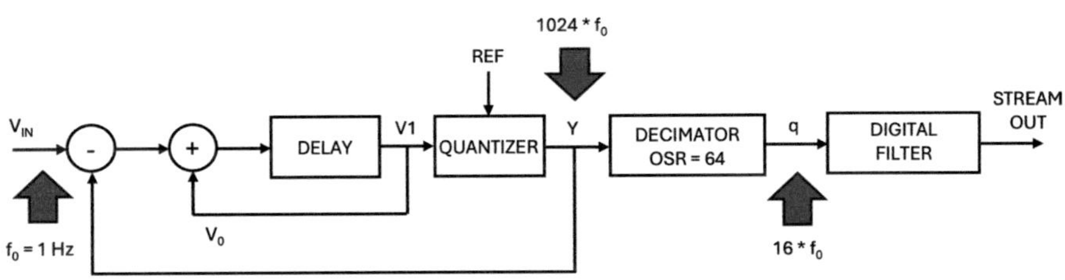

Figure 4.21 The sigma-delta quantization Matlab model [30].

indistinguishable. By contrast, the IIR filter implemented with the delta operator achieved a favorable balance (Table 4.7):

- It was less sensitive to SEUs, recovered from faults within a single cycle,
- Fewer flip-flops are required than the FIR design.

Quantitative comparisons showed that the delta-operator-based IIR filter (IIR Type 2) used fewer registers and multipliers than the FIR alternative, while being significantly more robust than the conventional IIR. For this ADC, any experimental result is available, too.

A further sigma-delta ADC is presented in [31]. The paper addresses the design of a sigma-delta analog-to-digital modulator intended for operation in extreme radiation environments, with a focus on space applications. The paper's authors identify two central challenges: radiation-induced threshold voltage shifts that can exceed 1 V, and the floating-body effects inherent in silicon-on-sapphire or partially depleted silicon-on-insulator CMOS technologies. The work aims to develop design techniques that enable mixed-signal circuits to maintain good performance under these hostile conditions. A multibit modulator architecture is chosen, not only for its resilience to radiation but also to serve as a vehicle for studying a variety of analog building blocks, such as switched-capacitor integrators, comparators, and flash converters. Ultimately, the goal was to demonstrate that with careful architectural and circuit design strategies (RHBD approach), it is possible to achieve radiation hardness beyond 20 Mrad (Si), while retaining high-resolution conversion suitable for demanding missions.

The proposed study begins with a careful analysis of floating-body effects in partially depleted SOI devices, which introduce drain current anomalies known as the kink effect, i.e. a sudden increase in the MOS drain

Table 4.7 Comparison between the three digital filter implementations [30].

Filter implementations	FIR	IIR Type 1	IIR Type 2
Register	21	6	6
Multipliers	22	15	15
Adders	21	9	15
Bits per register	8	16	13
Flip-flops	168	96	78

current. The authors propose circuit-level techniques, such as a new cascoding topology and radiation-hardened biasing, to suppress these variations. Specialized compact models for floating-body devices are employed in simulation to accurately predict behavior under bias and irradiation. In parallel, the paper investigates total-dose radiation effects, focusing on threshold voltage shifts that vary with device bias conditions. A custom preprocessor linked to the circuit simulator allows iterative modeling of device degradation as radiation dose accumulates. Circuit techniques such as chopper stabilization are introduced to mitigate radiation-induced input offset drift, particularly in comparators and differential amplifiers. For sampling, boosted-clock P-channel switches are adopted to avoid OFF-state leakage issues in irradiated N-channel devices. At the architectural level, a first-order multibit ΣΔ modulator with a 4-bit internal flash quantizer (Figure 4.22) was selected.

The design incorporates chopping at multiple points, differential resistor ladder DACs, and radiation-hardened op-amps and comparators. All key circuit blocks, including op-amps, comparators, and DACs, are specifically adapted to the floating-body SOS process.

The modulator was fabricated in a 1.5 μm space-qualified SOS CMOS process. The circuit achieved a dynamic range equivalent to 9.7 bits within a 63-kHz signal bandwidth. When subjected to radiation testing using a Co-60

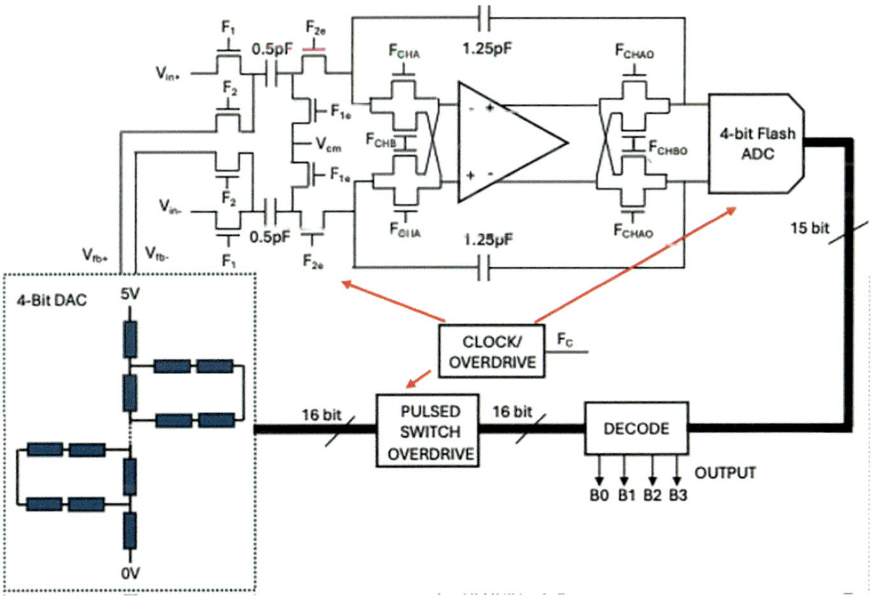

Figure 4.22 First-order multibit A/D architecture [31].

source, the devices were exposed up to 23 Mrad (Si), a dose level well beyond typical mission requirements. The dynamic range only degraded slightly, to 9.1 bits, even at this extreme dose. Detailed spectral measurements confirmed that noise shaping remained intact after irradiation, with only minor increases in distortion due to DAC mismatches. Simulations and experimental data demonstrated that the radiation-hardened cascode bias network maintained stable operating points under dose accumulation, in contrast to conventional cascode circuits that failed at high doses. Tests on comparators showed that chopper stabilization effectively suppressed large radiation-induced offsets. Sampling switches using boosted-clock P-channel devices maintained low ON and OFF resistances despite radiation exposure. A full set of irradiation experiments confirmed that all key functional blocks continued to operate as designed, with no catastrophic failures.

Beyond the demonstration of a single converter, the paper highlights a set of reusable analog building blocks that can serve as a foundation for other radiation-hardened mixed-signal systems. The study underscores that careful modeling, robust biasing, and architectural choices allow analog and mixed-signal circuits to be realized in SOS/SOI CMOS despite severe floating-body effects and extreme radiation environments.

A final example of rad-hard sigma-delta is proposed by [32]. The paper presents the design of a radiation-hardened delta-sigma analog-to-digital converter in 130 nm CMOS for use in extreme environments such as nuclear reactors, particle accelerators, and space missions. The circuit targets a 17-bit resolution over a 1 kHz bandwidth while consuming only 4.35 mW at a 1 MHz sampling rate. A correlated double sampling technique is employed to intrinsically suppress offset and 1/*f* noise, both of which increase under radiation. The architecture uses switched-capacitor integrators, a single-bit quantizer, and radiation-aware transmission gates to ensure robustness. SPICE-level simulations confirm an SNDR above 108 dB and stable performance across 0–85 °C. Radiation tests, modeled with threshold voltage shifts up to 1.36 MGy, show the ADC maintains its 17-bit resolution with only minor performance fluctuations. The design achieves a figure of merit (FOM) of 159, comparable to or exceeding state-of-the-art solutions. This paper demonstrates that careful circuit techniques and layout strategies can yield reliable, high-resolution ADCs for harsh radiation and thermal environments.

Successive approximation register (SAR) rad-hard ADC

A single event upset may arise either in the control logic or as a transient disturbance in the comparator, producing a bit-flip error during the conversion process. Because the switch states in later steps depend on the outcomes of the previous bits, an error in one bit can propagate through the sequence, leading to multiple erroneous bits in the result.

As the successive approximation register (SAR) operates sequentially, the earlier in the conversion an SEU occurs, the greater the likelihood that several bits of the output will be incorrect [10]. The same authors propose some first circuital solutions [33]. This work investigates the vulnerability of charge-redistribution successive approximation register analog-to-digital converters to radiation-induced single-event transients (SETs). The main objective is to analyze how transient faults in analog switches and digital control logic affect the conversion process, potentially leading to single or multiple bit errors. A second objective is to evaluate and compare mitigation techniques, focusing on transistor sizing, redundancy in control logic, and fault detection mechanisms such as built-in current sensors (BICS). The goal is to identify trade-offs between fault resilience, silicon area, power consumption, and conversion linearity, offering practical guidelines for radiation-hardened SAR ADC design. The SAR ADC under study follows the classical charge redistribution topology (see above), consisting of a capacitor array, comparator, switching network, and control logic. The capacitor array is binary weighted, while transmission gates and NMOS devices form the switches that connect capacitors to input, reference, or ground. The control logic is based on Anderson's shift register-based design, generating the correct switching sequence for each conversion step.

To model SET effects, the authors use SPICE-level simulations with a 130 nm predictive technology model. Faults are injected into sensitive nodes using a double-exponential current pulse model, which replicates the effect of an ion strike on semiconductor junctions. Parameters such as peak current, rise time, and fall time are chosen according to worst-case literature data. The campaign involves random fault injections across all conversion steps and nodes, covering both analog switches and digital control logic. This campaign shows that SAR converters are significantly affected by SETs. In the nominal design with standard transistor dimensions, around 22% of injected faults on analog switches caused conversion errors (Figure 4.23). The likelihood of errors was highest for least significant bits (LSBs), since smaller capacitors are more sensitive to imbalance. Similarly, faults in digital control logic produced errors in about 20% of cases, highlighting its vulnerability. The error distribution revealed that a single transient could propagate to multiple bits, especially when occurring in early conversion stages. This reflects the sequential nature of SAR operation, where an error in higher bits affects subsequent decisions. Quantitative data confirmed that most bit flips accumulated in the lower-order outputs, with error rates up to 26% in the LSB.

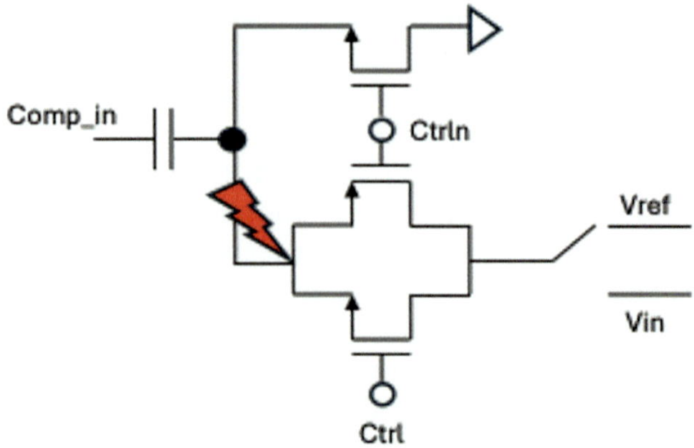

Figure 4.23 Schematic of one of the switches showing its most sensitive node [33].

Mitigation strategies are evaluated in three categories. First, transistor sizing in switches is studied: wider devices can tolerate higher charge without flipping states, but increase parasitic capacitances that degrade linearity. Second, redundancy approaches such as triple modular redundancy (TMR) are considered for the digital logic, although their area cost is high. Third, detection-based strategies such as bulk built-in current sensors are explored. These sensors monitor substrate currents to identify SETs and can reset the control logic to restart conversions, thereby avoiding propagation of corrupted results.

Experiments with different transistor widths showed a clear reduction in fault rates as device dimensions increased. Enlarging switch width reduced SET sensitivity but introduced trade-offs in parasitic capacitance, leading to potential non-linearities and charge injection issues. The optimal region appeared around moderate width increases, where robustness improved without excessive degradation of circuit performance.

The BICS approach was demonstrated through simulations as an effective detection mechanism. By resetting the control logic upon detection of a transient current pulse, faulty conversions were aborted and replaced by new samples. This technique added only minor delays, at most two sampling periods, and significantly improved the reliability of the conversion results.

On the other hand, Ref. [11] proposes a combination of circuit techniques, topology choices, and layout strategies to maximize radiation hardness without sacrificing performance or efficiency for a 10-bit ADC in a standard 0.8 μm CMOS process.

The proposed design methodology first addresses TID effects. Deep submicron technologies and ELTs are recommended, as they nearly cancel out TID-induced threshold shifts. However, these techniques pose challenges such as low supply voltages, stricter design rules, and potential matching issues in ELTs. Topology selection is also critical: converters should minimize the number of comparators, since comparators are highly sensitive to radiation-induced offsets, such as in SAR. A key technique introduced is autozeroing, which compensates for inevitable radiation-induced offsets. Two methods are presented: analog autozeroing, where offsets are canceled inside the comparator by extra circuitry and additional clock phases; and digital autozeroing, where oversampling allows offsets to be quantified and subtracted digitally from the ADC output. The Authors show how a digitally autozeroed successive approximation ADC can be built with two DACs sharing a resistive string, a comparator, a state machine, and a subtractor circuit (Figure 4.24). The digital autozeroing process involves two sequential conversions – one capturing the input plus offset, the other measuring the offset itself – so that subtraction cancels the error.

The DACs are resistor-based (resistors are insensitive to dose), while switch leakage does not affect the accuracy too much. To mitigate SEEs, the methodology advocates the use of large flip-flops in state machines, guard-rings, full-custom layouts for good P–N separation, and watchdog counters to reset the ADC in case of persistent SEUs.

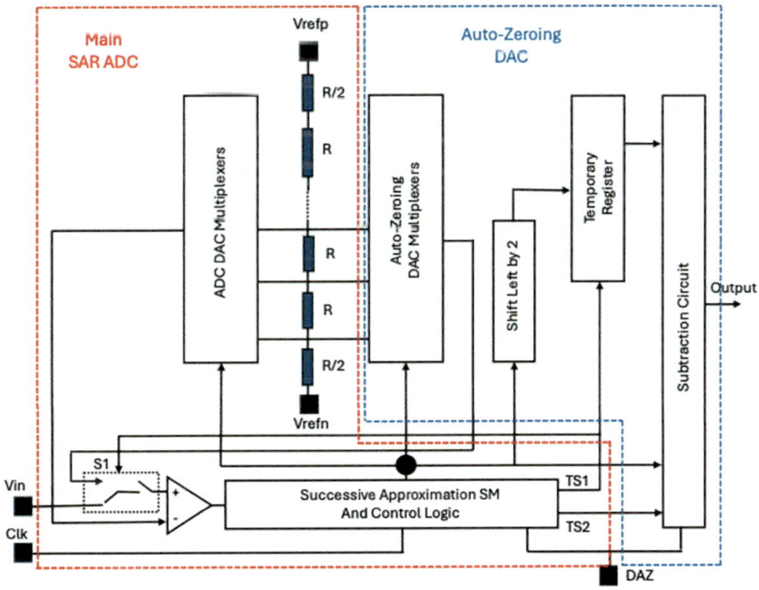

Figure 4.24 Proposed rad-hard successive digitally autozeroed SAR ADC [11].

The auxiliary auto-zeroing DAC (AZ-DAC) performs offset reduction as follows:

- Step 1: The analogue input is converted by the successive approximation ADC to a digital word (OC1), shifted one position left, stored to a temporary memory location, and fed to the AZ-DAC
- Step 2: The analogue input is isolated and the output of the AZ-DAC, which is the result of the first ADC, is converted to a digital word (OC2) by the successive approximation ADC
- When the second analog-to-digital conversion is completed, the result comes from the subtraction:

$$OC = 2*OC1 - OC2 \tag{4.2}$$

- Since it can be demonstrated that $OC1 = OC_{ideal} + offset$ and $OC2 = OC_{ideal} + 2*offset$, in a first approximation $OC = OC_{ideal}$

Experimental results up to 500 krad achieved the following performance (Table 4.8):

Another example of "analog" redundancy used to overcome errors induced by radiation in ADC is proposed by [34]. The paper aims to develop a high-resolution, radiation- and metastability-tolerant successive approximation register ADC for use in harsh environments such as the ATLAS experiment at CERN. These environments require ADCs that operate at 12–14 bits of resolution and speeds of 40–160 MS/s, while also withstanding radiation effects like single-event effects and total ionizing dose. Conventional SAR ADCs, although efficient, are vulnerable to metastability and radiation-induced upsets, which can corrupt data or cause catastrophic failures. The target is to design a 14-bit, low-power ADC with built-in resilience against radiation

Table 4.8 Measured performance of SAR ADC in [11].

Parameter	**Value**
ENOB without offset compensation	7 bit
ENOB with offset compensation	between 8 and 9 bit
Max conversion frequency	40 kS/s
INL (with OC)	<3LSB
Power consumption	<2 mW

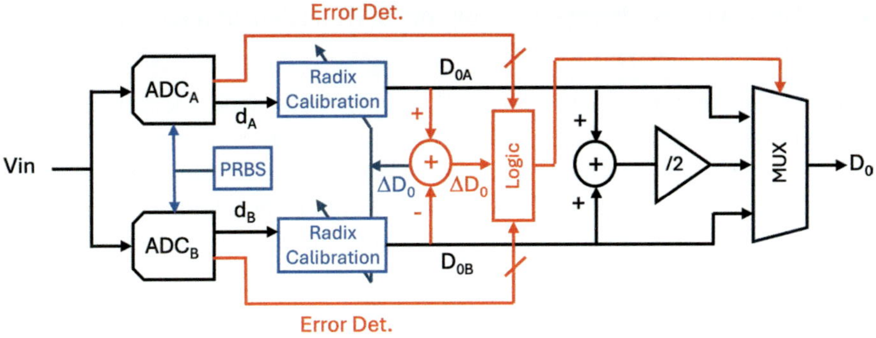

Figure 4.25 Prototype radiation-tolerant SAR ADC architecture [34].

and metastability, without the large overhead of traditional redundancy methods like TMR. To achieve this, the proposed ADC adopts a two-step split SAR architecture, which combines double modular redundancy (DMR) with error detection circuits. Two sub-ADCs process the same input in parallel; their outputs are compared to detect discrepancies. If both results are correct, averaging improves SNR; if one is wrong, it is discarded (Figure 4.25).

Within each sub-ADC, three specialized detection techniques are employed (Figure 4.26):

- Over-/under-flow (OU) detection – identifies abnormal residue voltages caused by radiation or metastability.
- Extra-bit (EB) cycle detection – uses an oversized capacitor in the LSB cycle to expose inconsistencies in the summing node.

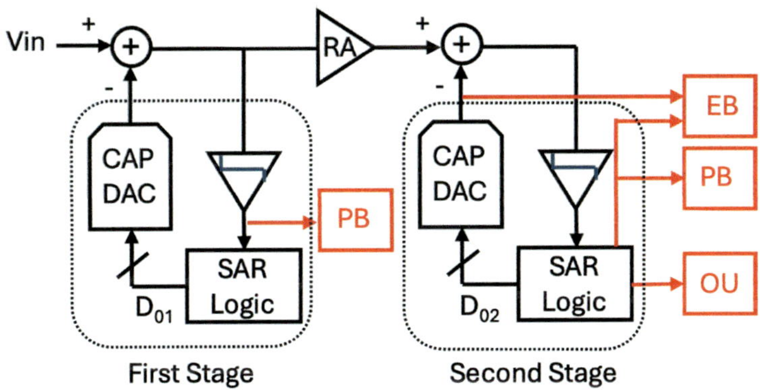

Figure 4.26 Embedded error detectors in the two-step SAR sub-ADC [34].

- Parity-bit (PB) detection – monitors SAR latch data integrity, ensuring bit flips are caught.

Architectural redundancy is reinforced by intra- and inter-stage redundancies through sub-radix-2 DACs and scaled reference voltages. The offset between the residue amplifier (RA) and comparator is mitigated using a shared pre-amplifier and calibration DACs. Background digital calibration corrects DAC mismatch using least-mean-square (LMS) algorithms, enabled by differences in transfer curves between the two sub-ADCs. On-chip reference buffers stabilize supply variations, improving DAC settling without large area penalties.

The converter has been fabricated in a 65-nm CMOS process. Measurements show good static linearity with DNL within ±0.42 LSB and INL within ±0.56 LSB at 14 bits. Dynamic performance achieved 78.5 dB SNDR and 103 dB SFDR at 35 MS/s, 78 dB SNDR and 95 dB SFDR at 40 MS/s, and >70 dB SNDR with >88 dB SFDR at 75 MS/s. Power consumption scales modestly from 22.2 mW at 40 MS/s to 24.9 mW at 75 MS/s, with most power used by amplifiers and reference buffers; redundancy circuits add only ~14% overhead.

Radiation testing validated robustness. Under a proton beam, large conversion errors were uncorrelated between the two sub-ADCs, and all were successfully detected and corrected, achieving a 100% correction rate. In X-ray TID tests up to 1 Mrad, SNDR (signal-to-noise and distortion ratio) degraded by less than 0.5 dB, SFDR by 2.3 dB, and power varied less than 0.5 mW. These results exceed ATLAS's 10-year radiation requirement of 75 krad.

The 12-bit SAR ADC presented in [35] acts mainly on the loop DAC to improve resiliency, particularly single-event transients (SETs). Prior research showed that although some radiation-hardening by design techniques improve robustness, they still leave the smallest capacitor node vulnerable. So, they proposed a design that maintains the basic SAR-ADC structure of control logic, SAR module, comparator, and charge redistribution capacitive DAC. The digital modules are designed in Verilog-HDL, while analog blocks are modeled with Verilog-A® for comparator tuning. To counteract the high susceptibility of the smallest capacitors, the Authors propose adding redundancy into the DAC and comparator structures. Two alternative redundant architectures are considered (Figure 4.27):

- Structure (a): Uses dual DACs and comparators, where if outputs disagree, the SAR process stalls until agreement is reached. This structure provides better fault tolerance and higher speed.
- Structure (b): Requires strict equality of outputs from two DACs before bits are accepted, offering improved reliability but at lower speed.

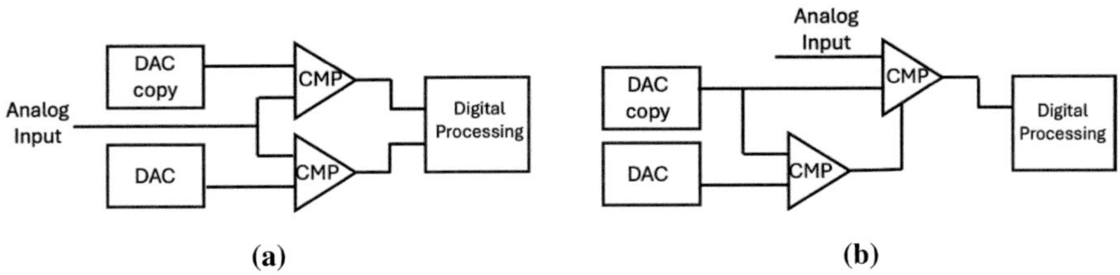

Figure 4.27 Redundant structure (a) and structure (b) in [35].

A "determiner" circuit checks comparator outputs to detect SET-induced mismatches. For the digital core, triple modular redundancy is applied to D flip-flops to mitigate single-event upsets. The design is implemented and simulated in 180 nm CMOS technology.

Fault injection simulations were performed using current source models to replicate ion strikes, with peak current pulses of 2 mA injected into the smallest capacitors of the binary weighted capacitive array of the DAC (C0, C1, C2). A test bench compared three cases: a standard SAR ADC without faults, a standard SAR ADC with injected faults, and the hardened ADC under the same conditions. Simulation waveforms showed that while the unhardened ADC produced erroneous outputs under injected SETs, the radiation-hardened design maintained correct conversions. The hardened ADC produced outputs identical to the fault-free reference, with minor Z-states used as markers for error detection. Two representative 12-bit outputs were analyzed: in both cases, the fault-injected standard ADC produced incorrect bit patterns, while the hardened ADC outputs matched the reference exactly. This confirmed the ability of the redundant DAC-comparator structure and hardened digital logic to suppress or mask SET-induced errors. However, the ADC has not been integrated, and only simulation results are available.

The work in Ref. [36] presents the design of a 12-bit SAR ADC with 1 MS/s sampling rate, less than 15 mW of power consumption, intended for the front-end readout of a 32-channel cadmium zinc telluride (CZT) detector imaging system. Besides suitable electrical performance, it has radiation tolerance. In its design, special emphasis is placed on ensuring offset-free comparator operation, reducing capacitor mismatch, and enhancing radiation tolerance through both circuit and layout strategies.

Starting from conventional SAR architecture, a hybrid DAC is used in the loop, consisting of a 6-bit charge-redistribution DAC (CDAC) combined with a 6-bit resistor-string DAC (RDAC) (Figure 4.28). This approach optimizes precision, monotonicity, and die area. To address the comparator offset, a modified output-offset-storage (OOS) technique is proposed. Unlike

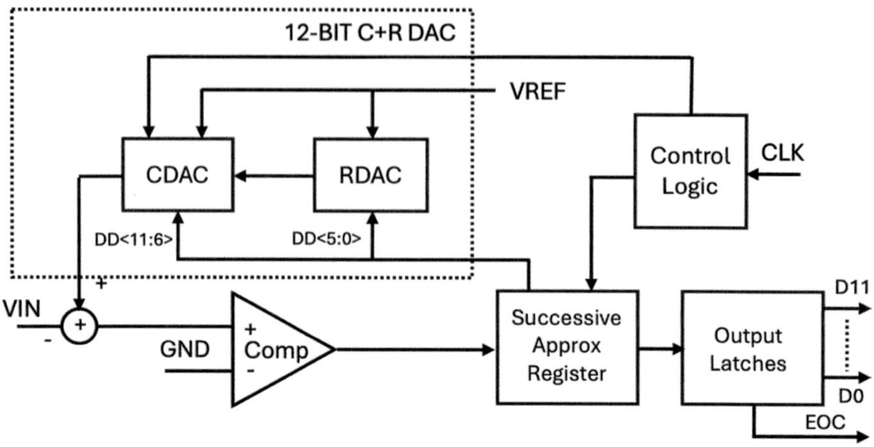

Figure 4.28 Specific architecture of proposed SAR-ADC [36].

conventional OOS, it avoids placing capacitors directly in the signal path, which improves speed and bandwidth while effectively canceling offset voltages that vary due to process or radiation effects. Capacitor mismatch is mitigated by using a unit capacitor array instead of binary-weighted capacitors. This reduces differential nonlinearity and integral nonlinearity, ensures monotonic transfer characteristics, and lowers switching currents, which in turn boosts speed.

Radiation hardening is incorporated at multiple levels. At the circuit level, comparator offset cancellation counters TID-induced shifts, while DSP-based redundancy filters help mitigate SEE-induced single-event upsets. At the layout level, guard rings and increased spacing between PMOS and NMOS devices are used to prevent latch-up. Enlarging transistor separation to 5 μm proved especially effective in suppressing SEL.

The prototype ADC was fabricated using TSMC 0.35 μm CMOS technology, occupying a die area of 1.18×1.08 mm^2. It operates with a 3.3/5 V supply and consumes only 10 mW at 1 MS/s. Static linearity tests using code density methods showed DNL ranging from –0.49 LSB to +1.34 LSB, and INL from –1.48 LSB to +1.1 LSB. Dynamic tests with FFT analysis yielded a SINAD of 67.64 dB, corresponding to an effective number of bits (ENOB) of about 10.94, with spurious-free dynamic range (SFDR) of 82.25 dB and total harmonic distortion (THD) of –77.98 dB. Power analysis indicated 6 mW static and 4 mW dynamic consumption, with most power drawn by the comparator and RDAC circuits. The ADC achieved a figure of merit (FoM) of 5.08 pJ/conversion-step. The SAR ADC has been integrated successfully with the front-end readout circuits. However, any testing for TID and SEE resilience is reported.

In paper [37], the authors therefore set out to design a low-overhead RHBD SAR-ADC with an error correction mechanism based on adaptive topology transformation (ATT) against SET. The proposed SAR-ADC achieves 13-bit resolution using differential input signals processed through dual capacitive DACs (C-DACs). A key innovation is the integration of radiation detectors near the sampling switches of each differential input. When a heavy ion strike produces electron-hole pairs that may inject extra charge into a capacitor, the detectors generate flag signals indicating the affected side. The system then adaptively transforms its topology: the comparator input switches from differential operation to single-ended operation, disconnecting the corrupted C-DAC and instead referencing the unaffected side against a stable common-mode voltage (Figure 4.29). This mechanism prevents the faulty charge from propagating through the approximation sequence.

The circuit includes a comparator with multiplexer capability to support this transformation, and a flag generator to process radiation detector signals. Analog blocks such as the C-DACs are implemented using bulk CMOS transistors at 3.3 V, while digital control circuits use SOI transistors at 1.2 V for inherent SEE resistance. The hybrid combination of RHBD techniques

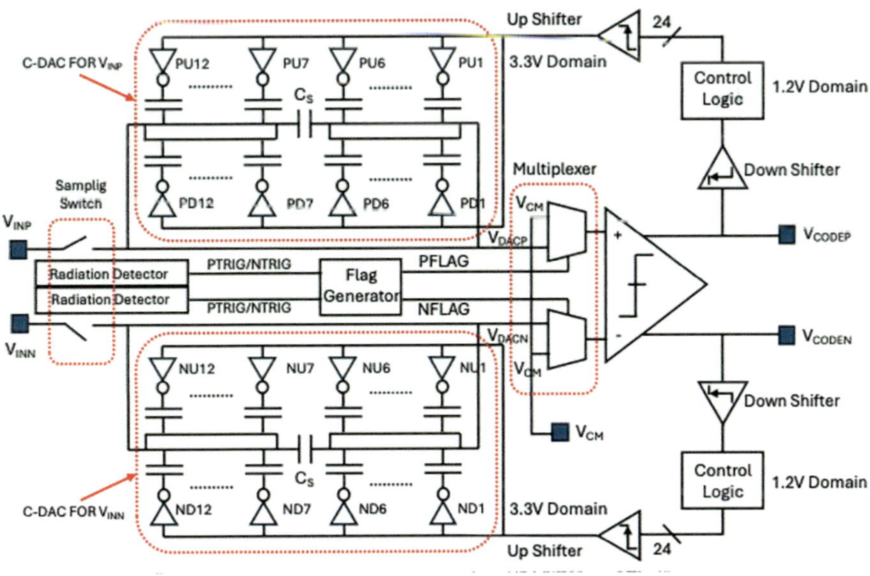

Figure 4.29 Circuit diagram of the proposed SAR-ADC [37].

for analog parts and RHBP (Radiation Hardened By Process) for digital parts optimizes resilience while keeping costs low. A test-chip fabricated in a Renesas 65 nm thin-BOX FDSOI (Fully Depleted SOI) process included eight variations of SAR-ADCs for comparative evaluation, with differences in control logic, detector sizing, and comparator configuration.

First, simulations showed that without protection, a charge injection event shifted digital output codes significantly from the expected value. With the ATT mechanism enabled, the ADC produced the correct code, confirming its effectiveness in mitigating SET-induced errors. For instance, when the correct digital output was 2604, charge injection alone caused a deviation to 2667, but the proposed design restored the correct value of 2604.

Then, measurements have been carried out under heavy ion irradiation (84Kr17+) at Tohoku University's CYRIC facility. The proposed SAR-ADC demonstrated marked resilience compared to a conventional baseline. At a common-mode voltage of 1.25 V, the conventional design showed its first error within 39 seconds, while the proposed design operated error-free for over 10 minutes under similar flux. Cross-section calculations revealed nearly an order of magnitude improvement at 1.25 and 1.45 V conditions. In terms of SNDR, the proposed ADC improved performance by more than 10 dB under flux levels above 10^6 count/cm^2 with respect to conventional.

However, although the design suppressed most error propagation, measurements showed a slight increase in noise floor compared to conventional SAR-ADCs, attributed to the comparator's topology. Moreover, due to the transformation from differential to single-ended mode during correction, the SNR theoretically decreases by about 6 dB. Nonetheless, the error reduction benefit outweighs this drawback in harsh environments.

The paper [38] is focused on circuital level solutions to improve tolerance to radiation without incurring large penalties in area or power, as well. The primary objective is to integrate a dual-capacitor sample-and-hold (S/H) mechanism and a TID-compensated digital comparator into a SAR ADC to enhance soft-error resilience. Thus, to strengthen resilience, the proposed design introduces three key modifications:

- First, a resistor-based DAC is used instead of a capacitor-based one, reducing sensitivity to SEE.
- Second, the SAR logic employs delay-based dual-feedback flip-flops to mitigate single-event upsets.
- Third, the comparator is implemented as a digital block with TID-monitoring and compensation circuitry, ensuring stable performance even under long-term irradiation.

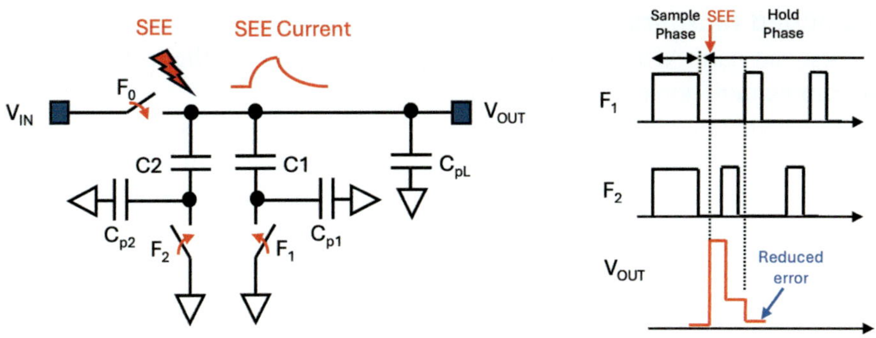

Figure 4.30 Schematic of the proposed dual-capacitor SAH (a); timing (b) [38].

The most innovative feature is the dual-capacitor sample-and-hold control. Instead of relying on a single capacitor, the input is sampled across two capacitors, C1 and C2, with staggered timing control (Figure 4.30). During the hold phase, these capacitors alternately compensate for charge disturbances caused by SEE, thereby reducing voltage error. This method provides error cancellation without significantly enlarging the circuit area. The design was implemented in 65 nm CMOS technology, with a core area of 0.17 mm^2. A test chip was fabricated and subjected to radiation exposure experiments using both alpha particle sources and high-intensity gamma-ray facilities.

The fabricated SAR ADC was evaluated under both laboratory and radiation conditions. Initial tests confirmed correct digital functionality, achieving a resolution of 10 bits with a measured ENOB of 8.01. The power consumption was around 210 μW. Radiation tests showed promising results. Under gamma-ray exposure up to 5 Mrad (Si), the ADC maintained stable ENOB values. Measurements of the dual-capacitor S/H confirmed that it significantly reduced output errors caused by SEE.

Further interesting proposals of rad-hard SAR ADC are in Ref. [39] and [40], where resiliency is achieved by replacing MOS with Dummy Gate Assisted (DGA), n-MOSFETs. These devices overcome the weaknesses of ELT, such as W/L ratio, complex W/L ratio modeling, an asymmetrical layout, and large gate capacitance, while supporting up to 500 krad (Si) dose.

Other solutions can be found in Ref. [41] and [42], where radiation-hardened-by-design techniques were applied, such as a custom digital standard cell library, using annular gate devices for NMOS layouts, guard rings around NMOS and PMOS transistors. Additionally, transistor spacing was increased to mitigate latch-up, and the active region was enlarged to reduce parasitic resistance. Redundancy against SEE is exploited also in Ref. [43] and Ref. [44].

Pipelined rad-hard ADC

A first rad-hard ADC proposed in [45] relies mainly on the technology to overcome radiation effects. The goal is to design a 12-bit, 3 GS/s, low-power ADC capable of tolerating high levels of total ionizing dose and single-event effects. Their main objective is to exploit the advantages of silicon-germanium (SiGe) heterojunction bipolar transistors (HBTs) and BiCMOS processes to achieve suitable performance in resolution, sampling rate, and radiation hardness. The architecture uses a time-interleaved (TI) structure with a full-speed front-end track-and-hold amplifier (THA) based on SiGe devices. This THA minimizes clock jitter problems by sampling the signal once at very low jitter, then distributing constant signals to slower pipeline channels. Each pipeline ADC channel processes data at lower rates, easing design complexity and avoiding skew calibration. Careful attention is given to the design of the high-gain, wide-bandwidth operational transconductance amplifiers (OTAs) for the 2.5-bit front stage, using only radiation-tolerant SiGe HBTs and PMOS transistors to avoid leakage issues found in thick-oxide NMOS (Figure 4.31). The chosen SiGe process (IBM 8HP, 130 nm) inherently withstands multi-megarad TID. Radiation simulations, including TCAD® modeling of heavy-ion strikes, were used to identify sensitive points, especially in the bandgap reference circuit, which was redesigned to reduce single-event transient durations. Calibration circuitry hosted in an external FPGA compensates for mismatches between interleaved channels.

All in all, provisions for rad-hard are:

- The inherent TID radiation tolerance of the 130 nm SiGe BiCMOS fabrication process is utilized to achieve hardness up to 2 Mrad (Si)

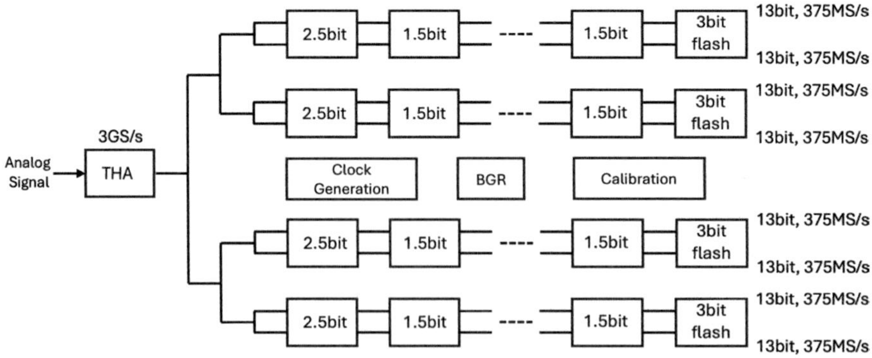

Figure 4.31 Time-interleaved pipelined ADC topology [45].

- In the amplifiers to avoid the radiation induced leakage currents, thick-oxide N-MOS transistors are omitted and only P-MOS transistors and SiGe HBTs are used

Only simulation results are available. They are summarized in Table 4.9.

Radiation testing data confirmed the intrinsic tolerance of SiGe HBTs to extreme TID levels (>100 Mrad) without parameter shifts. CMOS transistors within the BiCMOS platform showed vulnerability, but careful device selection avoided degradation. For single-event effects, long transients were initially seen in the bandgap reference, but redesign reduced them to about 275 ns.

An example of a pipelined ADC where resiliency is accomplished by means of circuital RHBD can be found in Ref. [9]. The objective of the authors is to identify weak points in a pipelined ADC, quantify their vulnerability, and propose hardening techniques to improve resilience. The test vehicle is a 10-bit conventional pipelined ADC built in a 130 nm CMOS process, operating with nine stages each resolving 1.5 bits. Digital error correction (DEC) was integrated to manage comparator offsets. To study SEE effects, the Spice-like simulator authors used the along with a Monte Carlo methodology. Over 25,000 simulations per linear energy transfer (LET) level were conducted, randomizing input voltage, strike time, and transistor location. Only the first stage was directly simulated, since its contribution dominates the output and its design repeats in later stages. Simulated LETs ranged from 1 to 20 MeV·cm^2/mg. Performance metrics such as transient amplitude and sensitive cross-section were extracted. Following the SEE analysis, for the analog part, the hardening technique has been focused on:

- the comparators of the 2-comparators sub-ADC
- the subtraction circuits

Table 4.9 Simulation results of TI Pipelined ADC in [45].

Parameter	Value
Resolution	13
Sampling rate	3 GS/s
SFDR@1.5 GHz	73 dB
ENOB	11.1 Bit
Power consumption	1.9 W
TID	2 Mrad
SEL	120 MeV·cm^2/mg

Thus, in the paper, two hardening methods were proposed: enlarging capacitors in the comparator subtraction circuit and resistive decoupling in comparator latches. Indeed, the comparator second stage is a latch implemented with a standard Static RAM (SRAM) design composed of two cross-coupled inverters. A single event in an SRAM that flips one side will flip the opposite side due to the positive feedback: if a delay is introduced between the two inverters, it is possible for the inverter that is hit to recover before the other inverter changes state, and thus prevent the upset. A common way is to use resistors to partially decouple the inverters. Further correction of comparator errors is delegated to the digital error correction (DEC) block. For the digital, pipelined triple-modular redundancy protects the digital output registers.

The application of these hardening techniques resulted in a reduction of the comparator sensitivity to single events with minimal impact on the circuit performance (mainly speed). Together, these methods significantly lowered cross-sections at higher upset thresholds: in some cases, the sensitive area decreased by two orders of magnitude compared to the original design. Trade-offs included increased capacitor area (~23.5% larger) and slight speed penalties in the latch due to RC delays. However, these penalties were manageable, given the ADC's relatively modest analog bandwidth (<100 MHz).

In Ref. [46], a radiation-hardened 13-bit pipeline ADC fabricated in a 90 nm standard CMOS process was designed and tested. The specific goals were to achieve 13-bit nominal resolution, an effective number of bits around 10.5 bits in uncalibrated mode, and a sampling rate of 80 MS/s while ensuring energy efficiency below 0.7 pJ per conversion step. An equally important objective was to validate the converter's robustness against TID and SEL. The ADC was implemented as a baseline of the known pipeline architecture with multiple stages, digital correction logic, and a carefully optimized amplifier design. Architecture optimization was performed with proprietary tools to balance thermal noise, power dissipation, differential non-linearity, and capacitance. The selected amplifier used a folded-cascode input stage followed by a differential pair, stabilized through enhanced cascode compensation to improve bandwidth. The design was validated across 17 process corners, ensuring reliability under worst-case conditions.

For radiation hardening, several strategies were employed in the paper: digital correction to absorb comparator offset variations, the use of low V_{dsat} devices for greater voltage margins, current mirrors to mitigate threshold shifts, and P+ guard rings to dissipate charge carriers and suppress parasitic latch-up paths. Key transistors were laid out with enclosed geometry to block parasitic leakage channels.

Experimental validation was carried out using both TID and SEL campaigns. TID testing up to 100 krad (Si) was performed on five samples plus one reference. Performance was monitored at a 40 MHz sampling rate, though the test setup and PCB introduced an estimated loss of about 2 bits in effective resolution. Even with this degradation, the converter maintained around 10.3 ENOB, SNR values above 67 dB, and SFDR close to 70 dB across supply voltages from 1.1 to 1.3 V. Linearity tests confirmed stable DNL and INL within acceptable limits under radiation exposure.

For SEE testing, heavy-ion irradiation was applied with energies ranging from 1 to 70 MeV·cm^2/mg, up to an LET of 69.99 MeV·cm^2/mg and fluence of 10^7 ions/cm^2. Across all conditions, no latch-up events are observed, confirming the effectiveness of the guard-ring and enclosed-layout strategies. Reference voltage, SNR, THD, and ENOB measurements showed no radiation-induced degradation.

The work in [47] focuses on comparing the performance of bulk CMOS and silicon-on-insulator CMOS technologies when used to implement radiation-hardened ADCs for space applications. The goal of this study was to design a 10-bit, 25 MS/s pipelined ADC prototype, implement it in both bulk CMOS and SOI CMOS processes, and experimentally compare the electrical performance and radiation tolerance. The ADC prototype adopted a pipeline structure with a front-end sample-and-hold circuit, three identical stages each performing 2.5-bit coarse quantization, and a final 4-bit flash stage. Digital error correction logic then generated the final 10-bit resolution (Figure 4.32).

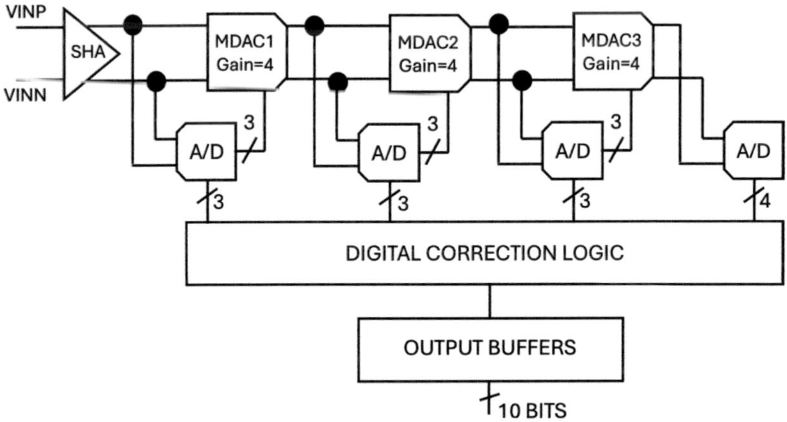

Figure 4.32 Block diagram of the ADC prototype [47].

To improve resilience, specific RHBD techniques were used: a capacitor flip-around sample-and-hold for higher bandwidth and lower power, a gain-boosted folded cascode op-amp for stability under TID stress, and switched-capacitor comparators with preamplifiers for lower offset and tolerance to radiation-induced errors.

In parallel, radiation-hardened-by-process methods were exploited by implementing the same design in SOI CMOS, which inherently reduces parasitic capacitance, eliminates latch-up, and provides smaller charge collection volumes, thereby improving SEE tolerance. For bulk CMOS, careful layout hardening techniques such as guard rings and substrate contacts were applied to mitigate latch-up risks.

The electrical performance tests showed that both SOI and bulk ADCs achieved similar baseline characteristics: ENOB around 9.5–9.7 bits, SNR of ~60 dB, SFDR ~72–74 dB, and DNL/INL within ±0.5 LSB at 25 MS/s. However, when the sampling frequency was increased, the SOI ADC maintained 8.2 ENOB up to 40 MS/s, while the bulk ADC degraded rapidly, dropping below 8 bits at 31 MS/s and to only 2.5 bits at 40 MS/s.

In TID experiments up to 500 krad (Si), the SOI ADC showed negligible degradation, maintaining stable consumption and spectral parameters, with tolerance confirmed above 300 krad (Si). By contrast, the bulk ADC began to fail after 50 krad (Si), losing effective resolution to ~4 bits at 100 krad (Si), though partial recovery occurred after annealing.

In SEE tests with heavy ions up to 63 MeV·cm^2/mg, neither design exhibited latch-up or functional interrupts (SEFI). However, the SOI ADC had a single-event upset cross-section two orders of magnitude lower than the bulk ADC: ~3.1E-6 cm^2/device at LET 17 MeV·cm^2/mg and ~9.6E-6 cm^2/device at LET 63 MeV·cm^2/mg. These smaller deviations were often corrected by the pipeline's digital error correction logic, further improving robustness.

A hybrid pipelined-SAR ADC is presented in Ref. [48]. Beyond the usual trade-offs of power, speed, and accuracy in ADCs, radiation tolerance emerges as a fourth critical factor. Traditional solutions such as TMR are costly in area and power. The objective of this study is to propose and demonstrate a 13-bit, 80 MS/s pipelined-SAR ADC in 65 nm CMOS that achieves high efficiency and strong radiation robustness, while minimizing penalties in area and power. The design is based on a semi-time-interleaved (Semi-TI) pipelined-SAR structure, in which two ADC channels share a single residue amplifier (RA). This improves efficiency by maximizing RA usage while also lowering per-channel operating speed, which benefits radiation tolerance.

To counteract TID-induced offsets, the RA employs ping-pong auto-zeroing with two alternating op-amps, reducing offset while minimizing power dissipation. The op-amp itself uses a fully differential folded-cascode class-AB architecture with gain boosting, laid out with an interleaved centroid for mismatch and SEE resilience. Biasing circuits were hardened using node-splitting and averaging techniques to mitigate charge collection from heavy ions. For clock generation, TMR was applied to frequency dividers, avoiding single-event functional interrupts. The comparator includes time-out protection through a dummy comparator and is split into smaller units for SEE hardening. Other measures include the use of DICE (Dual Interlocked Storage Cell) flip-flops, deep N-wells, guard rings, and limited minimum transistor dimensions. Overall, the design applies selective hardening to only the most sensitive sub-blocks to balance efficiency and robustness, thus avoiding severe area penalties.

Shown electrical tests demonstrate strong static and dynamic performance. The ADC achieves ±0.7 LSB DNL, ±1.38 LSB INL, and at 80 MS/s with a 38 MHz input reached 70.8 dB SNDR and 80.3 dB SFDR, while consuming only 13.8 mW from a 1.2 V supply.

Irradiation tests validated robustness. X-ray TID testing up to 500 krad (Si) showed stable SNDR, SFDR, and power consumption in both conversion and hot-spare modes, with no measurable degradation even after annealing. Laser-based SEE experiments indicated that most blocks recover within a single sample; vulnerable elements such as RA bias circuits and auto-zero switches showed limited sensitivity but did not compromise overall operation. Importantly, no latch-up was observed during testing.

Finally, we would like to mention a technique that involves only the architectural level. It is the so-called "digitally assisted techniques" (DAT) [49]. It is motivated by the rapid advancement of digital technology, which allows for improving an ADC (or a DAC) design by correcting or calibrating against static and dynamic degradation. In classical designs, they can be categorized into trimming of elements, foreground and background calibration, and dynamic element matching. The same approach can be applied to correct degradation due to radiation. Moreover, these techniques can be considered to a certain extent architecture-independent, a feature desirable for rad-hard data converters.

One of them is the so-called "multi-path ADC technique":

- In digital system design, triple modular redundancy is extensively used for control and data paths to provide radiation hardening against SEU

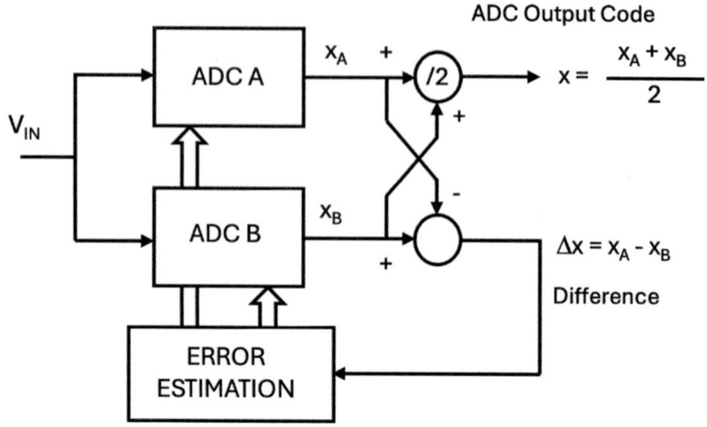

Figure 4.33 "Split ADC" architecture [50].

- The multi-path ADC technique is a mixed signal design equivalent to the digital TMR technique, which enables a robust correction method for single-event errors in an ADC
- The correction is done with a deterministic digital calibration procedure fully in the digital domain, with proven rad-hard libraries

This technique does not introduce penalties of analog area, power, and performance, because it is based on the "split ADC" approach [50].

In concrete, an ADC can be "split" as in Figure 4.33.

This approach is called a "split ADC" rather than a "dual ADC," because it divides the analog section of a single ADC design. The effect on analog complexity is minimal, with little influence on overall size, power usage, bandwidth, or noise. Figure 4.34(a) shows the chip area of a single

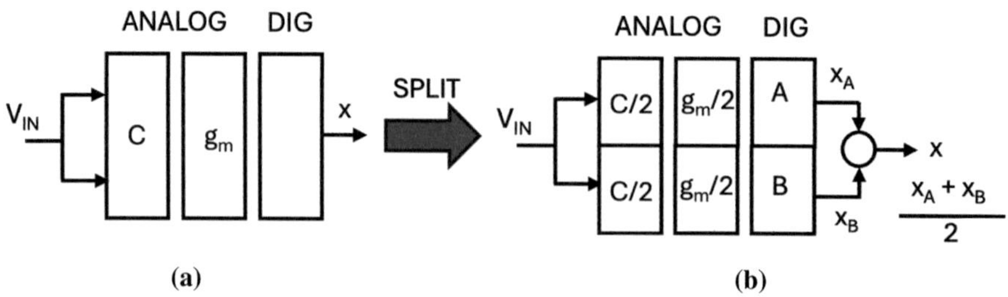

Figure 4.34 Effect of split on speed, power, noise, area [50].

N-bit ADC built to satisfy requirements for power, speed, and noise. Within this diagram, one portion represents the active analog blocks, another portion corresponds to passive parts like capacitors, and the rest is the digital logic. It is assumed that bandwidth is proportional to g_m/C, power to g_m, and noise to (KT/C)**0.5. The proportionality factors depend on circuit details, but they remain unchanged after splitting because the design is simply halved. Figure 4.34(b) illustrates the ADC divided into two equal converters, each occupying half of the original analog area. Each converter still achieves the same bandwidth and consumes the same total power as the original.

Although noise in each half rises by (2)**0.5, combining the two outputs by averaging cancels this effect, keeping the overall noise the same. The digital portion may become slightly more complex, but in modern CMOS processes, this is a minor increase and worth the trade-off.

When this concept is applied to a pipelined ADC [50] simulation results are as in Table 4.10.

The extra digital hardware needed for SEE correction in a multi-path ADC consists of the fractional interpolation filter in the two-path design and the majority (or minority) voting blocks in the three-path design. In the case of the two-path ADC, adding the SEE correction block improves the effective resolution from under 6 bits (without correction) to 12 bits, though only up to half of the Nyquist bandwidth, $F_s/4$. For the three-path ADC, the majority-voting correction approach restores full performance across the entire Nyquist bandwidth $F_s/2$. It is important to note that the correction precision and bandwidth achieved by the three-path approach cannot be matched by interpolation-based methods. In fact, the accuracy of interpolation correction quickly degrades for input frequencies above $F_s/2$, making it unsuitable for Nyquist rate ADCs.

Table 4.10 Comparison of single, two-path, and three-path ADC performance [50].

Parameter	Conventional	Two-paths	Three-paths
SNR without SEE	12 bit	12 bit	12 bit
Core analog area	Same	Same	Same
Digital overhead	--	Interpolation and Digital Logic	Majority voting and digital logic
SEE correction	None	Method 1	Method 2
SNR with correction	< 6 bit	12 bit	11.72 bit

Ramp rad-hard ADC

The final class taken into consideration is the ramp ADC (single or dual-slope), mainly used in high-energy physics experiments and image sensors. An example can be found in [16]. In this work, the radiation robustness of a newly designed Wilkinson analog-to-digital converter implemented in the IBM 0.5 μm SiGe BiCMOS 5HP process is investigated. This ADC forms part of a cold-capable, analog sense ASIC used within a distributed motor controller (DMC) intended for next-generation Mars rover missions. The objective is to evaluate how the Wilkinson ADC tolerates total ionizing dose, proton irradiation, and single-event effects, and to identify its most sensitive circuit elements to guide hardening strategies.

The studied device is a 16-channel, 12-bit Wilkinson ADC with less than 5 mW of power consumption, designed for precision sensing of motor health parameters such as temperature and stress. Its topology uses a shared ramp generator and counter across channels, each with three auto-zeroing comparators, a latch, and a register. The input stage includes transmission gates (TGs), which proved central to the SEE response.

Radiation evaluation combined several approaches:

- Broad-beam heavy-ion irradiation at Texas A&M with LETs up to 85.4 MeV·cm^2/mg,
- Pulsed-laser experiments at JPL (Jet Propulsion Laboratory) to map sensitive nodes,
- Mixed-mode simulations to replicate and analyze transient behaviors.

Three input channel configurations were tested: driven by sample-hold circuits, telemetry multiplexer outputs, or left undriven/open, to study the role of input impedance.

Cumulative radiation tests (X-ray and proton up to 300 krad) showed negligible degradation in DNL, SNR, or power, confirming SiGe's strong TID resistance. No single-event latch-up (SEL) was observed even under high LET and overvoltage.

SEE tests revealed sensitivity to single-event upsets strongly dependent on input configuration. Channels actively driven showed no SEUs, while undriven or high-impedance channels experienced long single-event transients, with cross-sections increasing with LET. Laser experiments confirmed that strikes on the PMOS transistors of the transmission gates reproduced the heavy-ion-induced long SETs. Simulations further demonstrated that the input impedance network governs the duration of SET pulses: low impedance quickly recovers, but high impedance prolongs the disturbance, corrupting digital outputs. Importantly, the TMR counter proved immune to SEUs, validating the mitigation strategy. Overall, the ADC remained functional, but its SEE sensitivity was highly input-dependent, particularly in undriven channels.

Other rad-hard ramp ADCs can be found in [14] and [15].

4.4 0.6 MS/S 12-bit SAR Rad-hard ADC

Starting from the above overview and related considerations, the SAR solution seems the most popular and promising in terms of simplicity, consumption, and resiliency. So, the Authors developed and verified a 0.6 MS/s ADC having 12-Bit as embedded interface in microcontrollers. The technology used was a 130-nm CMOS, as it provides sufficient integration density for mixed-signal circuits while ensuring manageable radiation hardness. The target performances of the prototypes are Total Dose immunity up to 300 krad (Si), resiliency to latch-up up to 60 MeV*mg/cm^{2,} and SEU of 10 MeV*mg/cm^2 for the digital controlling section.

The ADC adopts a conventional SAR structure with a charge-redistribution DAC. This choice offers a simple and efficient solution, minimizing ambiguity in separating radiation-induced effects from inherent ADC errors. SAR ADCs are widely used for medium-to-high resolution applications due to their compact area and low power, making them attractive for space systems. In this design, SEU and SET mitigation were applied to the SAR logic. Flip-flops were reinforced by adding capacitive elements to sensitive nodes, thereby increasing the charge required to upset logic states. The DAC was realized as a split binary-weighted capacitor array, with MIM capacitors stacked to increase reliability.

4.4.1 ADC architecture and circuit design

The proposed ADC employs a successive approximation register architecture with a charge-redistribution capacitive DAC (Figure 4.35). The structure is differential even if not shown in the figure. SAR converters are well-suited for designs where low power and compact area are essential. They sequentially resolve N bits in N clock cycles, offering higher efficiency than flash ADCs at medium-to-high resolutions, since flash architectures beyond 10 bits are impractical.

SAR's reliance on both combinational and sequential logic makes it a representative case study for understanding SEE and how to overcome it. Indeed, the logic in the proposed design uses enhanced D-type flip-flops (D-FF) with transmission-gate structures. To strengthen these circuits against

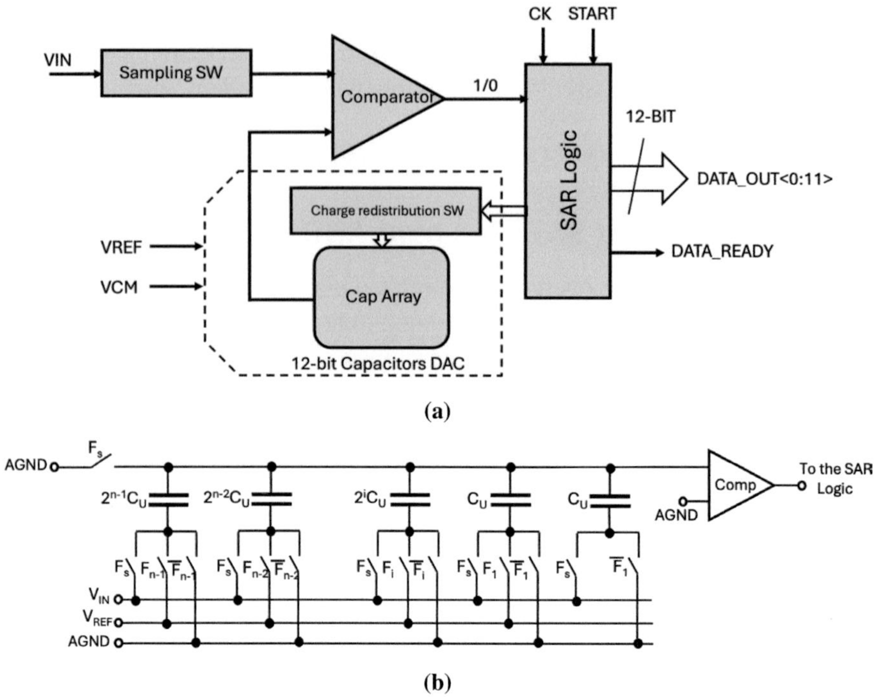

Figure 4.35 SAR ADC: Block diagram (a); charge redistribution DAC (b) [53].

SEEs, extra capacitances were added to the most vulnerable nodes (marked as "Logic-cap"), implemented using MOS devices (Figure 4.36). This raises the critical charge ($Q_{crit} = C_{node} \times \Delta V$), making it harder for an ion strike to

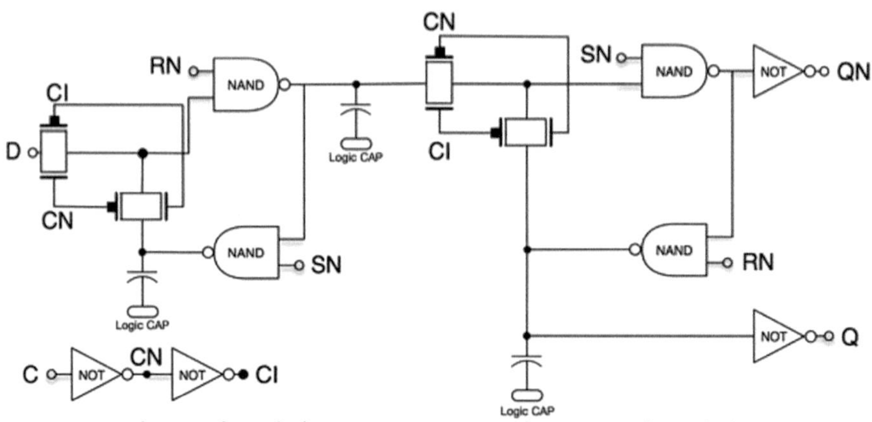

Figure 4.36 Rad-hard delay flip-flop [53].

induce an upset. By dimensioning PMOS and NMOS transistors appropriately, the circuit balances SEE resilience and speed.

On the other side, the DAC is implemented as a split binary-weighted capacitor array, with 7 bits in the most significant branch and 3 bits in the least significant one. The capacitors are realized using stacked metal-insulator-metal (MIM) devices, which are robust against dielectric damage from radiation. By using a unitary capacitor approach, the layout of the array is optimized in a common centroid topology to improve matching (Figure 4.37).

At the layout level, RHBD practices are essential for preventing SEL and leakage. Enclosed layout transistors are used extensively, removing parasitic paths at device edges. By arranging NMOS and PMOS transistors inside diffusion guard rings, leakage (inter-device) and latch-up are further suppressed. Although ELTs consume more area and their models are less accurate, they significantly improve TID and SEE resilience. Figure 4.38 illustrates the layout of the proposed rad-hard D-FF. An extensive use of ELT shapes guarantees resiliency to TID and to avoid SEL (latch-up), a careful guard-ring in layout has been used.

Switch dimensions in mixed-signal circuits require trade-offs between linearity, speed, and robustness, as done in DAC (see Chapter 3). In this design, complementary switches with ELT geometry were employed, as well, leading to wider transistors than the process minimum, but ensuring radiation tolerance. Switches robust layout is like that one adopted in the

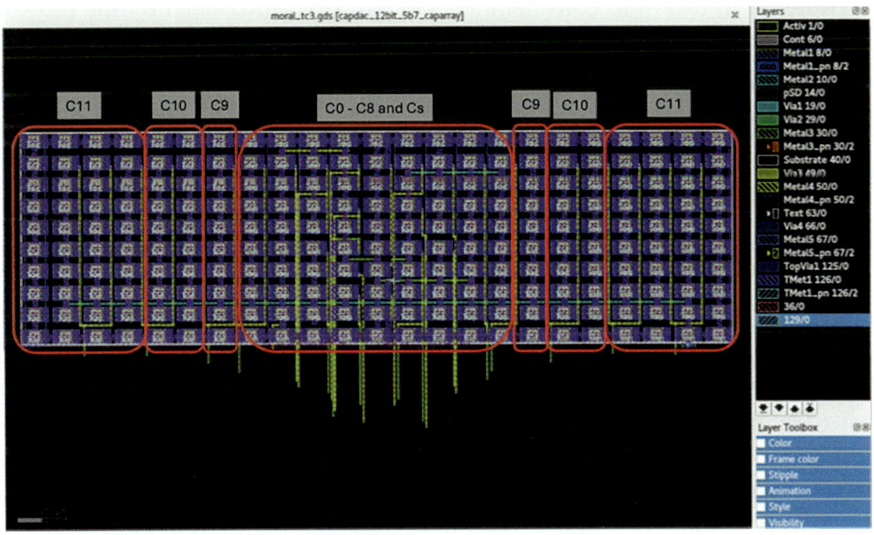

Figure 4.37 Layout of the unitary capacitors array with common centroid topology.

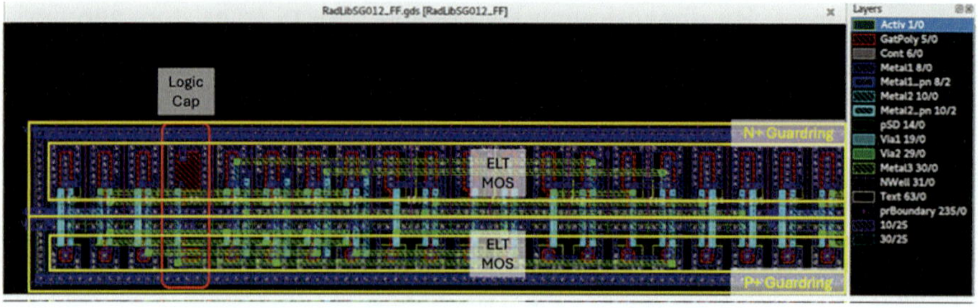

Figure 4.38 The layout of the proposed D-type flip-flop.

DAC. The comparator, being the most critical analog block, was designed as a preamplifier which incorporates a dynamic latch (Figure 4.39(a)). The preamplifier reduces offset and mitigates kickback noise, while the latch provides the necessary decision speed. In the physical layout, all the transistors with low length, *L*, will be designed as ELT, while those with higher length will be designed as regular, since the parasitic MOS appearing at transistor edges after radiation exposition is less effective. For the input differential pair (L = 1 um) and the current tail (L = 3 um). The designed comparator must be laid out considering especially the stringent requirements on offset. This implies the use of strong symmetrical structures; in particular, a common centroid topology is introduced in layout, which guarantees good matching of the differential stage in all directions. Extraction and post-layout simulations allowed us to take into consideration the parasitic introduced by the physical layout and mismatch performance. Guard rings were carefully applied in its layout to protect against SEE, as well. The input differential stage with inter-digitated common centroid layout for ELT is shown in Figure 4.39(b).

The D-FF and the control logic have been implemented with cells belonging to a rad-hard standard-cells library. The cell library, developed in comparable technologies, was previously characterized and shown to withstand up to 300 krad (Si) of TID, SEUs at 10 MeV·mg/cm^2, and SEL events at 60 MeV·mg/cm^2. The digital portion of the DAC (including control logic and decoders) was implemented using this library. In the overall ADC layout, extended guard rings (EGRs) were inserted wherever feasible to suppress latch-up, too.

General mixed-signal layout rules were also followed: separate analog and digital supplies, interdigitated device layouts, and multi-layer routing for low-resistance references.

The overall SAR ADC floor plan is shown in Figure 4.40(a). As can be seen, the structure will be fully-differential to mitigate common-mode

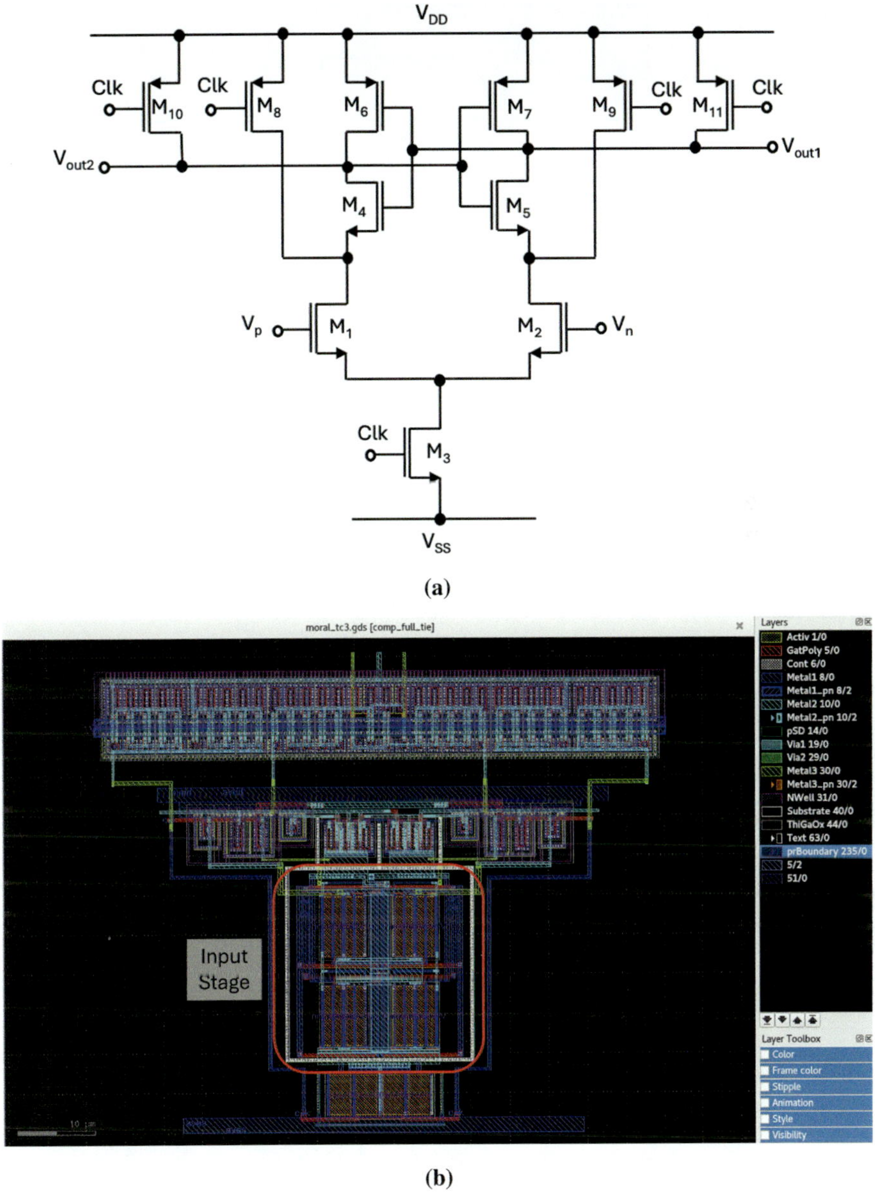

Figure 4.39 Schematic (a) and layout (b) of the comparator input differential stage with common-centroid ELT.

disturbances (noise, offset, etc.). The layout of the SAR ADC macro is shown in Figure 4.40(b). The size is 0.5 mm × 0.4 mm (0.2 mm^2).

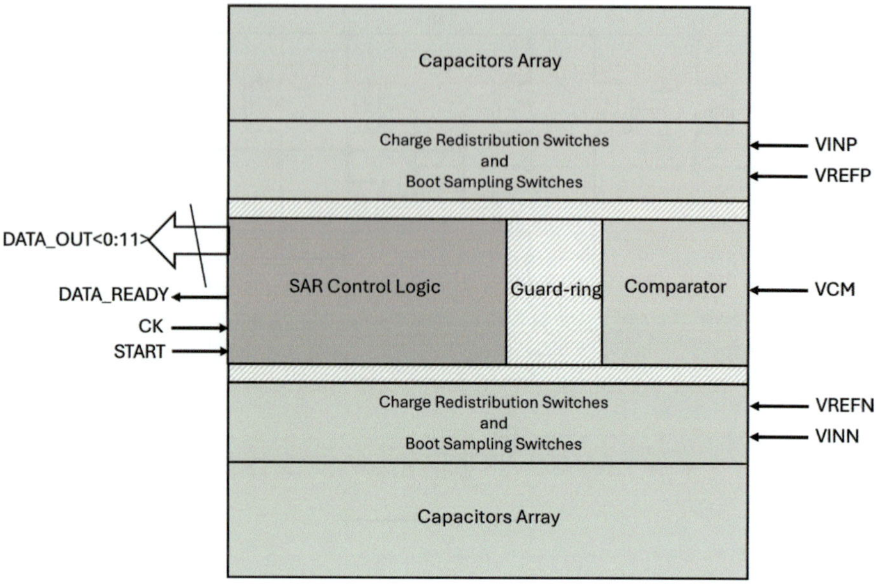

(a)

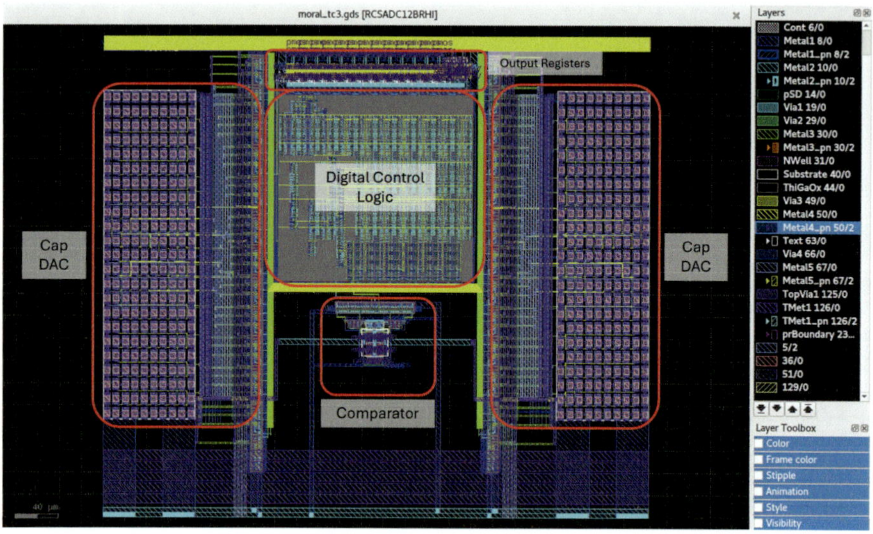

(b)

Figure 4.40 Floor plan (a) and layout of the 12-bit ADC macro (b).

The SAR ADC has been integrated in a standard 3.3 V 130-nm CMOS process (IHP SG13S). Figure 4.41 shows the micro-photo and the assembled component in a 32-pin CERDIP.

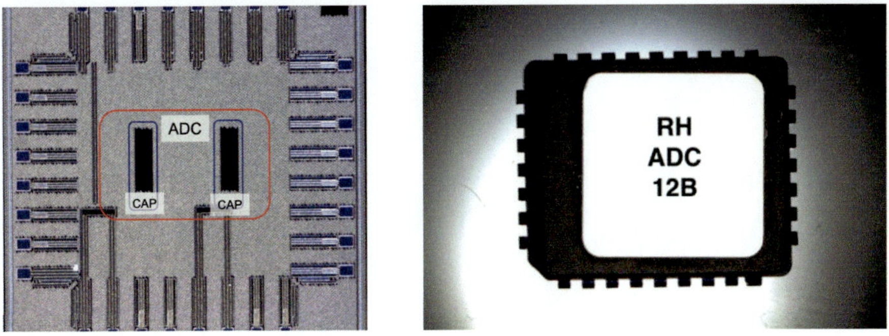

(a) (b)

Figure 4.41 Chip photo (a) and packaged ADC (b).

4.4.2 In-lab measurements

Both 10-bit and 12-bit prototypes have been preliminarily characterized in terms of static and dynamic performance. The output voltage range has been fixed at 1.5 V. The DNL and INL have been measured and are shown in Figures 4.42 and 4.43.

The measured 12-bit DNL is –1/+2 LSB, while INL is –6/+8 LSB.

The dynamic performance has been measured by applying a sinewave and using a clock of 3.3 MHz (0.2 MS/s) and then a clock of 10 MHz

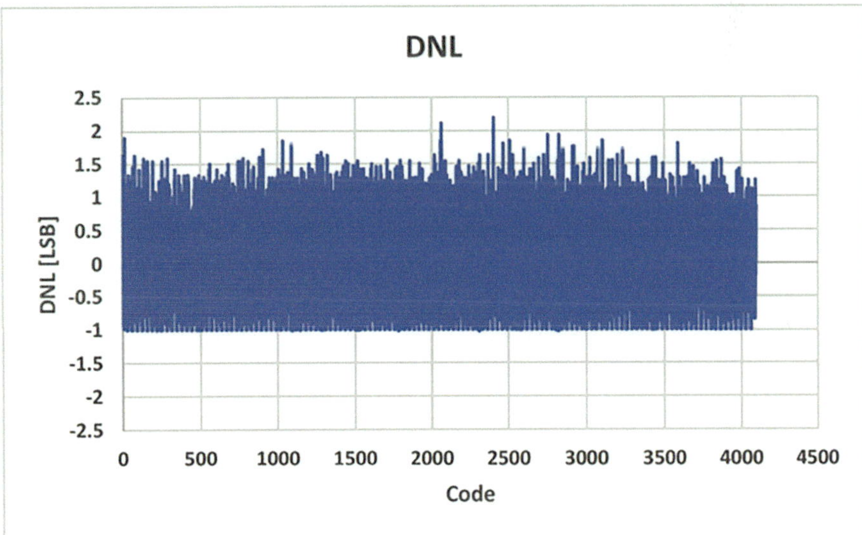

Figure 4.42 Measured DNL 12-bit ADC.

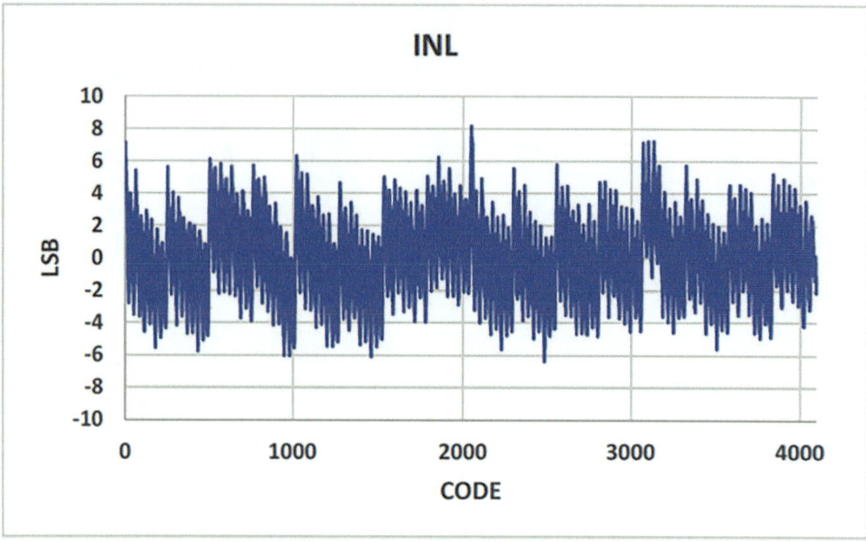

Figure 4.43 Measured INL of the 12-bit ADC.

(0.6 MS/s). The measured spectrum when a 3.3 MHz clock is applied is shown in Figure 4.44. ENOB accounts for noise and total harmonic distortion (THD). ENOB is higher than 9.6 bits.

In Figure 4.45, the spectrum of the same signal is shown when a 10 MHz clock is applied. ENOB is 9.5 bits.

Intermodulation (IMD) performance (IP3) has been measured, as well. A pair of tones equal to 250.549 kHz + 269.470 kHz is applied with a clock of 10 MHz; IMD is lower than 65 dBc (Figure 4.46).

Core power consumptions from a 3.3 V supply are in Table 4.11.

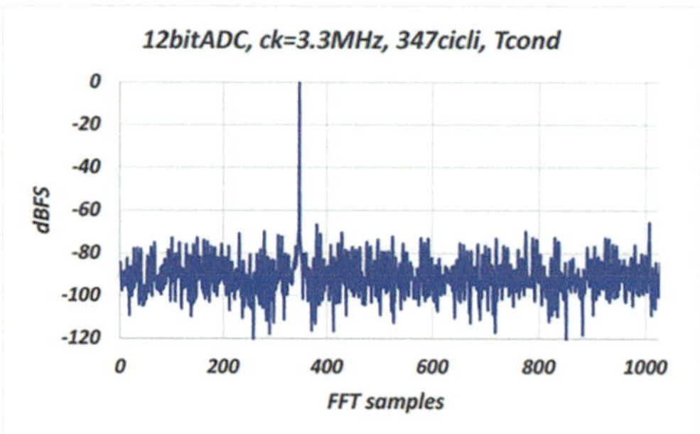

SFDR=	-65.00	dBFS
snr=	-61.50	dBFS
THD=	-64.20	dBFS
SINAD=	-59.60	dBFS
ENOB=	9.60	bits

Figure 4.44 Measured spectrum of the ADC output when a 35 kHz 2 $V_{pp,\,diff}$ sinewave with a clock of 3.3 MHz is applied.

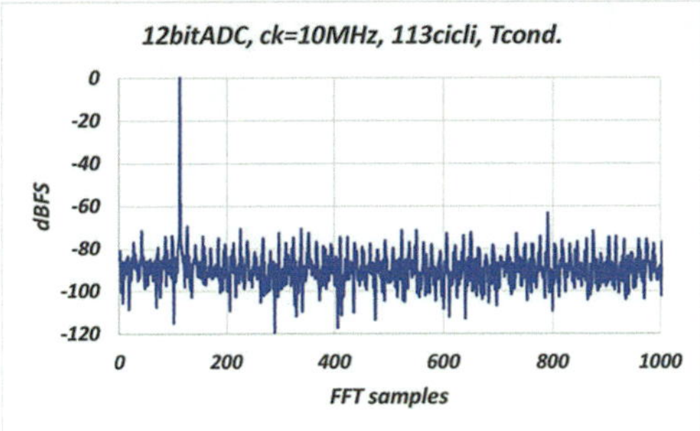

SFDR=	-64.00	dBFS
SNR=	-59.64	dBFS
THD=	-69.00	dBFS
SINAD=	-59.12	dBFS
ENOB=	9.50	bits

Figure 4.45 Measured spectrum of the ADC output when a 35 kHz 2 $V_{pp,\ diff}$ sinewave with a clock of 10 MHz is applied.

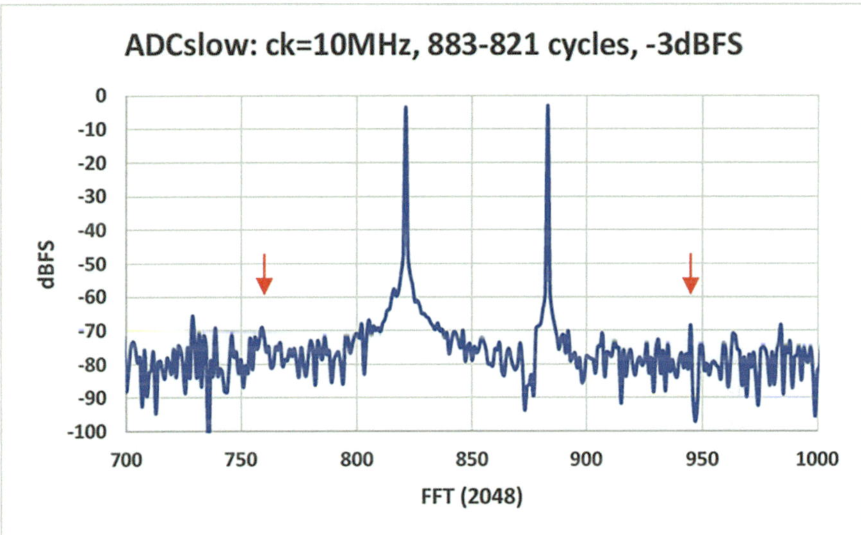

Figure 4.46 Measured IMD (IP3) when the buffered output of two tones with a clock of 10 MHz is applied (IMD = 65 dBc).

Table 4.11 Core power consumptions with a 3.3 V supply.

	Normal operation (10 MHz)	Power-down
12-bit ADC	<15 mW (of which 0.5 mW digital)*	<200 uW

*Mainly by pad circuitry (from simulation, comparator consumption is less than 1.5 mW)

4.4.3 Measurements under radiation

The rad-hard 12-bit ADC has been tested under ^{60}Co at the University of Palermo facility using the RedCat Devices proprietary Leonida Platform. Using a 0.63 rad/s dose rate, during the experiment, the ADC reached up to 400 krad (Si) with no significant degradation in terms of electrical performance.

Due to the extreme environmental conditions, especially the high levels of radiation, it is crucial to develop an appropriate testing methodology. Traditional methods – such as using a test board with the device under test (DUT) mounted and lab equipment to generate signals and record outputs – are not viable. Only the DUT should be subjected to radiation, while measurement tools and auxiliary circuits must remain outside the irradiation zone. As a result, all measurement procedures must be conducted remotely.

As done for the DAC, the proposed solution is based on splitting the test board into two boards: a mother-board (MB), which hosts all the controls, exposed, and a daughterboard (DB) with the DUT only. This approach is described by means of several examples in Chapter 5. Here, as the mother-board, an off-the-shelf Arduino Due board plus a custom shield is used.

The differential analog inputs of the proposed ADC are generated by very accurate DACs onboard the Arduino Due. The digital output of the proposed ADC is recorded by Arduino Due and further elaborated. In the following, the measurement results obtained under TID with such methodology are presented.

For the TID measurements, we sequentially applied a digital input code to the Arduino DAC (corresponding to 0, mid-scale, and full-scale values). It converts to an analog signal fed to the proposed SAR ADC. Then, the ADC output is recorded on the Arduino Due. The analog input range of the ADC is theoretically set from 0.9 to 2.5 V (mid value 1.65 V). Additionally, we measured the static power consumption – primarily attributable to the analog section – using a suitable current sensing circuit.

Figure 4.47(a) presents the SAR ADC output voltage when a mid-scale analog signal is applied as a function of the dose (from 0 to 350 krad (Si)). The absolute median values recorded (1.65 V) stay stable with TID. The small fluctuation is due to the inaccuracy of the measurement setup. Then, this test is repeated for the max and min output voltages (Figure 4.47(b,c)).

Then, an analog ramp and a sinewave are applied to the SAR ADC. Figures 4.48 and 4.49 represent the obtained results. Apart from measurement noise, the ADC output voltages remain unchanged with TID up to 350 krad (Si).

Any deviation from the core power consumption from the analog supply is observed up to 350 krad (Si).

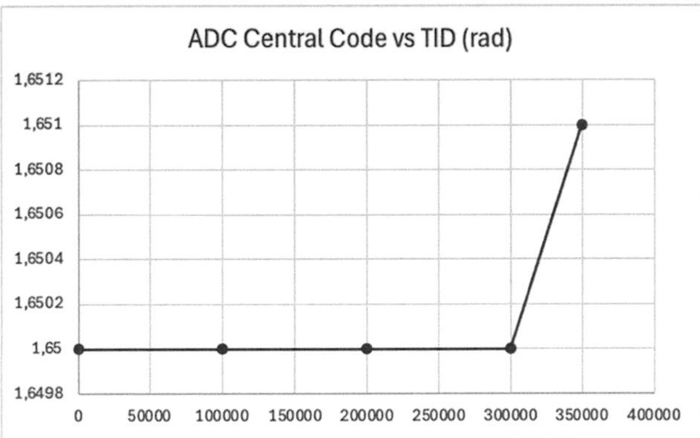

(a)

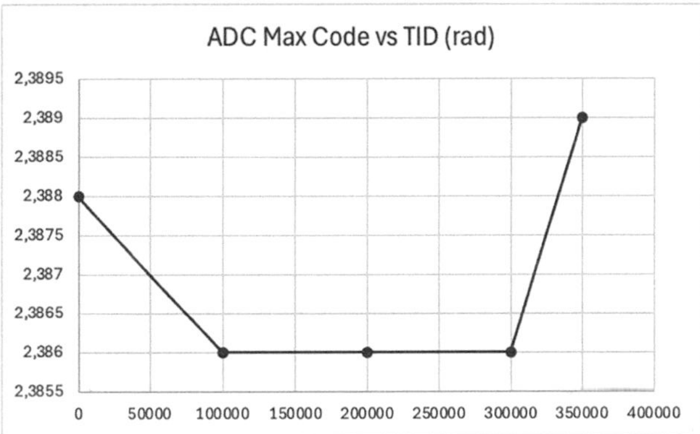

(b)

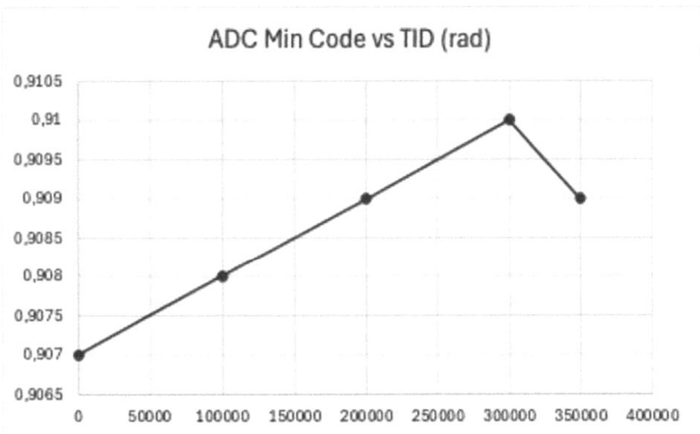

(c)

Figure 4.47 ADC output voltages as a function of the dose.

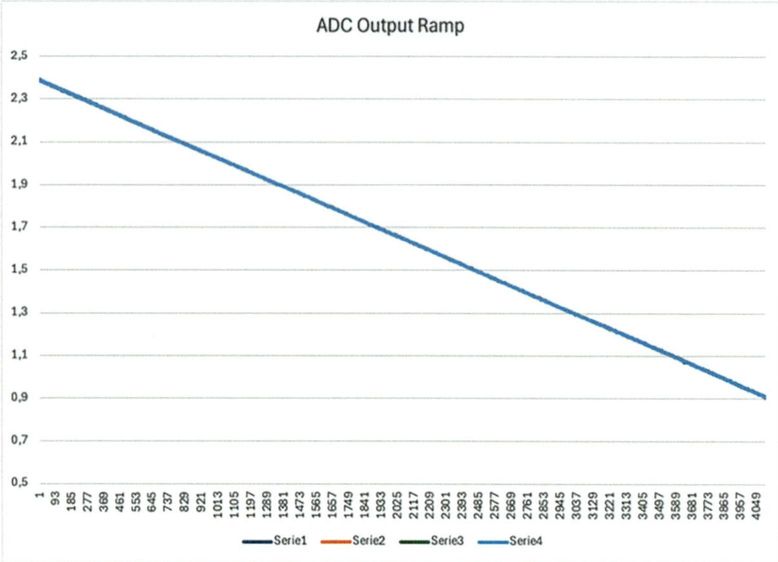

(a)

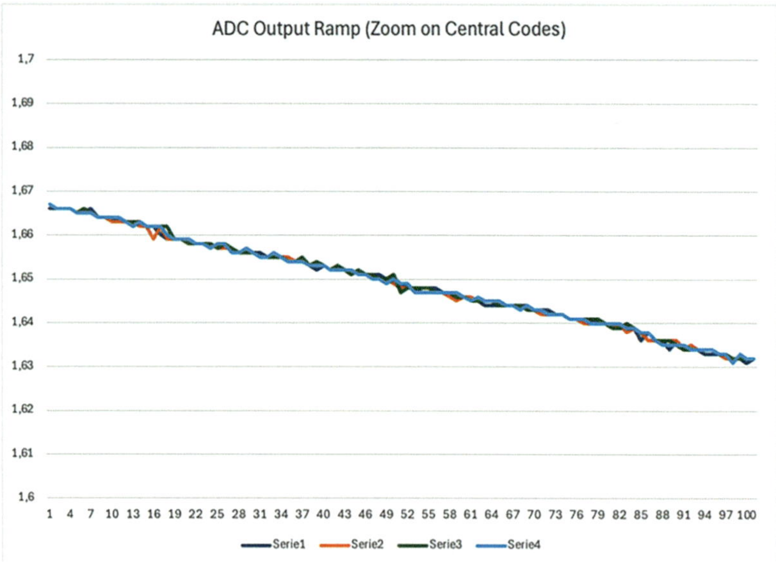

(b)

Figure 4.48 ADC output ramp: Serie1 = 0 rad (Si), Serie2 = 100 krad (Si), Serie3 =2 00 krad (Si), Serie4 = 400 krad (Si). Overall view (a) and zoomed view (b).

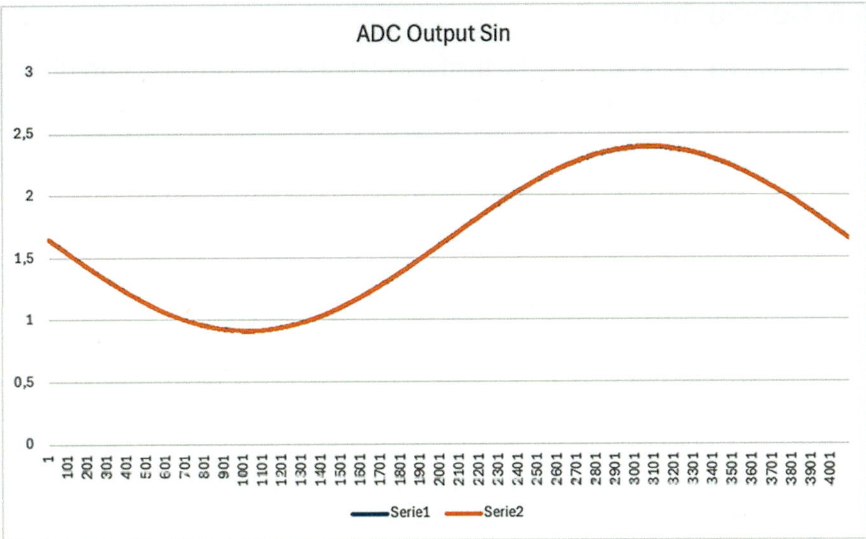

(a)

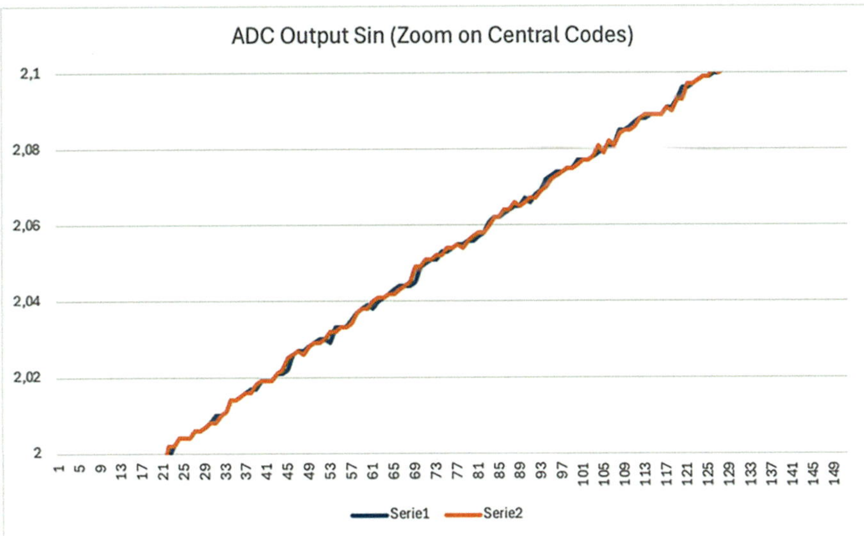

(b)

Figure 4.49 ADC output sinewave: Serie1 = 0 rad (Si), Serie2 = 100 krad (Si), Serie3 = 200 krad (Si), Serie4 = 400 krad (Si). Overall view (a) and zoomed view (b).

4.5 Conclusion

In this chapter, we examined the fundamental principles of analog-to-digital converters, which serve as key building blocks at the interface between the analog and digital domains and are frequently integrated as embedded components within SoCs. Proper design of this class of circuits is essential, as their characteristics directly influence the speed, dynamic range, and accuracy of the systems or interfaces in which they are employed. This becomes even more critical in harsh electronic environments, where, in addition to common degradation factors such as temperature fluctuations, supply variations, and device mismatches, radiation tolerance also becomes a decisive requirement. For this reason, we provided an overview of the main design strategies reported in the literature to address these challenges, supported by multiple examples, aiming to offer the reader both practical design insights and a comprehensive reference list covering recent advancements. Finally, we presented a radiation-hardened SAR-based ADC developed by the authors, which features total dose immunity up to 300 krad (Si), an estimated resiliency to latch-up up to 60 MeV*mg/cm^2, and SEU of 10 MeV*mg/cm^2 for the digital controlling section.

References

[1] F. Maloberti, "Analog Design for CMOS VLSI Systems", Springer, 2001.
[2] B. Razavi, "Principle of Data Conversion System Design", IEEE Press, 1995.
[3] R. Jacob Baker, "CMOS Circuit Design, Layout, And Simulation" R. Jacob Baker Ed.. IEEE Press, 2019.
[4] M. Pelgrom, "Analog-to-digital Conversion", Springer 2016.
[5] J.L. McCreary; P.R. Gray, "All-MOS charge redistribution analog-to-digital conversion techniques. I", IEEE Journal of Solid-State Circuits, Volume: 10, Issue: 6, December 1975, pp. 371-379.
[6] K. Karadamoglou, N. P. Paschalidis, et al., "An 11-bit High-Resolution and Adjustable-Range CMOS Time-to-Digital Converter for Space Science Instruments", IEEE Journal of Solid-State Circuits, Volume: 39, Issue: 1, January 2004, pp. 214-222.
[7] P. Leroux, et al., "Radiation-Tolerant Delta-Sigma Time-to-Digital Converters", Springer 2015.
[8] U. Gatti, C. Calligaro, E. Pikhay, Y. Roizin "Radiation-hardening methodologies for Flash ADC", Analog Integrated Circuits and Signal Processing Journal (Springer), DOI 10.1007/s10470-015-0627-7.

[9] A. L. Sternberg, L. W. Massengill, et al., "Single-Event Sensitivity and Hardening of a Pipelined Analog-to-Digital Converter", IEEE Transactions on Nuclear Science, Volume: 53, Issue: 6, December 2006, pp. 3532-3538.

[10] A. Lanot, T. Balen, "Analysis of the Effects of Single Event Transients on an SAR-ADC based on Charge Redistribution", 2014 15th Latin American Test Workshop – LATW, March 2014.

[11] G. Kottaras, N. Paschalidis, et al., "A 10 bit, low power, digitally auto-zeroed CMOS Analogue to Digital Converter Core for the NASA TRIO Smart Sensor System on a Chip", Proceedings of the 28th European Solid-State Circuits Conference, Sept. 2002.

[12] K. Makris, D. Fragopoulos, et al., "Very High Resolution Analog-to-Digital Converter at 1 kHz for Space Applications", Proc. AMICSA 2014, Zurich, July 2014.

[13] F. Bouyjou, O. Gevin, et al., "A 32-Channel 13-b ADC for Space Applications", IEEE Transactions on Nuclear Science, Volume: 64, Issue: 4, April 2017, pp. 1071-1079.

[14] F. Ciciriello, C. Marzocca, et al., "A Rad-hard 12-bit Auto-Calibrated ADC in CMOS 65nm", 2017 IEEE Nuclear Science Symposium and Medical Imaging Conference (NSS/MIC), Atlanta, Oct. 2017.

[15] T. Vergine, M. De Matteis, et al., "A 32-Channel 12-bits 65nm Wilkinson ADC for CMS Central Tracker", Proc. 2014 10th Conference on Ph.D. Research in Microelectronics and Electronics (PRIME), Grenoble, June 2014.

[16] P. C. Adell, J. Yager, et al., "Radiation Hardening of an SiGe BiCMOS Wilkinson ADC for Distributed Motor Controller Application", IEEE Transactions on Nuclear Science, Volume: 61, Issue: 3, June 2014, pp. 1236-1242.

[17] C. Calligaro, U. Gatti, et al., "Radiation-hardened techniques for CMOS flash ADC", 21st IEEE International Conference on Electronics, Circuits and Systems (ICECS), Marseille, Dec. 2014.

[18] Calligaro, C.; Arbat, A., et al., "A 15 Mrad (Si) 512Kbit Rad-hard SRAM in a standard 0.18um CMOS technology", Proceedings of the 3th European Conference on Radiation and its Effects on Components and Systems (RADECS) 2012, Biarritz, France, 24–28 Sept. 2012.

[19] C.I. Lee; A.H. Johnston, "Comparison of total dose responses on high resolution analog-to-digital converter technologies", IEEE Transactions on Nuclear Science, Volume: 45, Issue: 3, June 1998, pp. 1444-1449.

[20] T.L. Turflinger, et al., "Radiation effects in analog CMOS analog-to-digital converters", 1996 IEEE Radiation Effects Data Workshop. Workshop

Record. Held in conjunction with The IEEE Nuclear and Space Radiation Effects Conference, Indian Wells (CA), July 1996.

[21] E. Savasta, "First S-Band capable dual 12bit 1.5GSPS ADC in flip-chip hermetic technology", AMICSA 2016, Gothenburg, June 2016.

[22] P. Layton, et al., "TID Performance Degradation of High Precision, 16-bit Analog-to-Digital Converters", 2003 IEEE Radiation Effects Data Workshop, Monterey (CA), July 2003.

[23] A. Zanchi, et al., "Total Ionizing Dose and Single Event Latch-up Characterization of a 16-bit A-to-D Converter Fabricated in 0.18um Triple-Well Process", Proc. NSREC (Nuclear and Space Radiation Effects Conference), 2008, Miami, USA, pp.

[24] F. Baille, G. Borel, et al., "A Multi MRad hardened 8 bit/20 MHz flash ADC", IEEE Transactions on Nuclear Science, Volume: 39, Issue: 3, June 1992, pp. 401-404.

[25] R. Heuner; V. Zazzu; L. Pennisi, "Processing and circuit design enhance a data converter's radiation tolerance", IEEE Transactions on Nuclear Science, Volume: 35, Issue: 6, December 1988, pp. 1552-1556.

[26] N. van Vonno, "A radiation-tolerant 8-bit flash A/D converter implemented in silicon-on-sapphire", Proc. RADECS 91 First European Conference on Radiation and its Effects on Devices and Systems, La Grande-Motte, Sept. 1991.

[27] T. Monnier, et al., "Restructuring of the flash A/D converter to improve SEU rad tolerance", Proc. 1999 Fifth European Conference on Radiation and Its Effects on Components and Systems. RADECS 99, Fontevraud (France), Sept. 1999.

[28] A. Namazi, et al., "Highly reliable A/D converter using analog voting", 2008 IEEE International Conference on Computer Design, Lake Tahoe (CA), Oct. 2008.

[29] Wei Huang, et al., "Design of TID-Tolerant ADCs in Multivoltage Domain Based on Body Effect", IEEE Transactions on Nuclear Science, Volume: 70, Issue: 10, October 2023, pp. 2278-2284.

[30] F. P. Cortes, et al., "A ΣΔ A/D Converter Insensitive to SEU Effects", Proc. On-Line Testing Workshop, IEEE International, 2002.

[31] C. F. Edwards, et al., "A multibit Sigma Delta modulator in floating-body SOS/SOI CMOS for extreme radiation environments", IEEE Journal of Solid-State Circuits, Volume: 34, Issue: 7, July 1999, pp. 937-948.

[32] J. Verbeeck, "A radiation hard delta Sigma ADC in 130nm CMOS", Proc. PRIME 2012; 8th Conference on Ph.D. Research in Microelectronics & Electronics, Aachen, June 2012.

[33] A. J. C. Lanot, "Fault Mitigation Strategies for Single Event Transients on SAR Converters", Proc. 19th Annual International Mixed-Signals, Sensors, and Systems Test Workshop Proceedings, Porto Alegre, Sept. 2014.

[34] Hongda Xu, et al., "A 78.5dB-SNDR Radiation- and Metastability-Tolerant Two-Step Split SAR ADC Operating up to 75MS/s with 24.9mW Power Consumption in 65nm CMOS", Proc. 2017 IEEE International Solid-State Circuits Conference (ISSCC), Feb. 2017.

[35] H. Tang, et al., "A Radiation-Hardened 12-bit SAR ADC Design", Proc. 2018 7th International Conference on Reliability, Infocom Technologies and Optimization, Noida (India), Aug. 2018.

[36] Wei Liu, et al., "Design of a 12-bit 1 MS/s SAR-ADC for front-end readout of 32-channel CZT detector imaging system", Nuclear Instruments and Methods in Physics Research Section A: Accelerators, Spectrometers, Detectors and Associated Equipment, Volume 786, pages 155-163, 2015.

[37] Yuya Aoki, et al. ," A 13-bit Radiation-Hardened SAR-ADC with Error Correction by Adaptive Topology Transformation", Proc. 2023 IEEE International Reliability Physics Symposium (IRPS), Monterey (CA), March 2023.

[38] D. Ro, et al., "A Soft-Error-Tolerant SAR ADC with Dual-Capacitor Sample-and-Hold Control for Sensor Systems", Sensors (MDPI), in 2021, Volume 21, Issue 14, article number 4768.

[39] Tae Hyo Kim, Hee Chul Lee, "Dynamic Performance of a Radiation Tolerant 12B SAR ADC Designed with a DGA-MOSFET", Proc. 2013 IEEE Nuclear Science Symposium and Medical Imaging Conference (2013 NSS/MIC), Seoul, Oct. 2013.

[40] Tae Hyo Kim, Hee Chul Lee, "Total Ionizing Dose Effects on a 12-bit 40kS/s SAR ADC Designed With a Dummy Gate-Assisted n-MOS-FET", IEEE Transactions on Nuclear Science, Volume: 64, Issue: 1, January 2017, pp. 648-653.

[41] F. Xue, et al., "A Radiation-Hardened Low-Power Pipelined SAR ADC for CZT-based Imaging System", Proc. 2015 IEEE Nuclear Science Symposium and Medical Imaging Conference (NSS/MIC), San Diego (CA), Oct. 2015.

[42] A. Rivetti, et al., "A low-power 10 bit ADC in a 0.25 um CMOS: design considerations and test results", Proc. 2000 IEEE Nuclear Science Symposium, Lyon, Oct. 2000.

[43] Y. Chiu, "Radiation-Tolerant SAR ADC Architecture and Digital Calibration Techniques", Seminar at CERN EP-Electronics Seminars, 26 May 2015.

[44] H. Xu, et al., "High-speed, high-resolution, radiation-tolerant SAR ADCs for particle physics experiments", Journal of Instrumentation (IOP Publishing), Volume 10, Issue 04, April 2015.

[45] E. O. Mikkola, "Ultra-low-power radiation hard ADC for particle detector readout applications", Journal of Instrumentation (JINST), Volume 8, Issue 04 (April 2013).

[46] B. Vaz, et al., "Design and Testing of a Radiation Hardened 13-bit 80 MS/s Pipeline ADC Implemented in a 90nm Standard CMOS Process", Second International Workshop on Analog and Mixed Signal Integrated Circuits for Space Applications (AMICSA 2008), Sintra, Portugal, Sept. 2008.

[47] F. Qi, et al., "Comparison Study of Bulk and SOI CMOS Technologies based Rad-hard ADCs in Space", Workshop on Analog and Mixed Signal Integrated Circuits for Space Applications (AMICSA 2016), Gothenburg, Sweden, June 2016.

[48] Zheyi Li, et al., "An 80 MS/s 70.8 dB-SNDR Radiation-Tolerant Semi-Time-Interleaved Pipelined-SAR ADC for Space Applications", IEEE Transactions on Circuits and Systems I: Regular Papers, Volume: 72, Issue: 8, Aug. 2025, pp. 3897-3910.

[49] H. Venkatram, et al., "Detection and Correction Methods for Single Event Effects in Analog to Digital Converters", IEEE Transactions on Circuits and Systems I: Regular Papers, Volume: 60, Issue: 12, Dec. 2013, pp. 3163-3172.

[50] J. McNeill, et al., "Split ADC" Architecture for Deterministic Digital Background Calibration of a 16-bit 1-MS/s ADC", IEEE Journal of Solid-State Circuits, Volume: 40, Issue: 12, Dec. 2005, pp. 2437-2445.

[51] P. Bellandi, et al., "A 12-Bit Pipelined Self-Calibrated Radiation Hardened ADC for Aerospace Applications", RADECS 2002 (Radiation and its Effects on Components and Systems), Padova, Italy, Sept. 2002, pp. 141–144.

[52] P. R. Fleming, et al., "Design Technique for Mitigation of Soft Errors in Differential Switched-Capacitor Circuits", IEEE Transactions on Circuits and Systems II: Express Briefs, Volume: 55, Issue: 9, Sept. 2008, pp. 838-842.

[53] M. Baghbanmanesh, et al., "A 10-Bit Radiation-Hardened by Design (RHBD) SAR ADC for Space Applications", Proc. Conference 2017 New Generation of CAS (NGCAS), Genova, Sept. 2017.

5

Testing of Rad-hard Analog Devices under Radiations

Abstract

In this chapter, we describe the technologies and testing methodology developed for evaluating electronic devices under radiation exposure. The test setup consists of irradiation facilities (gamma rays for TID and Cyclotron beam for SEE), dedicated hardware boards and fixtures, and supporting software tools. The approach is sufficiently general to enable the verification of various classes of electronic components. In this chapter, first, the basic concepts of testing under radiations are presented, and as a simple example, we report measurements performed on SRAM modules, but the same procedure has been readily applied to analog devices such as DACs, ADCs, voltage regulators. The performance metrics obtained range from bit error counts and power consumption in memories to linearity (SNR, THD) and resolution (ENOB) in DACs and ADCs, to absolute output voltage for voltage regulators, confirming the effectiveness of the proposed methodology. Three significant examples of test on the rad-hard components developed by the authors and described in the previous chapters are described in detail. The output of their testing is the measurement results already presented.

5.1 Introduction

The requirements for TID and SEE vary widely depending on the intended application of the device. In space missions, for instance, the total ionizing dose absorbed by silicon typically spans from a few krad(Si) up to several hundred krad(Si), depending on factors such as the spacecraft's mission profile, orbital parameters, and the level of shielding used [1, 2]. For example, in the case of low orbital satellites, the required TID is ~100 krad [3]. For devices employed in interplanetary missions or in high-energy physics

DOI: 10.1201/9788743813286-5

experiments, the exposure is far greater, with TID levels ranging from a few Mrad(Si) to several Mrad(Si) [4, 5].

Thus, it is very important to realize irradiation methodology and test facilities for electronic devices simulating the different radiation conditions to which they may be prone, the kind of operations and the intrinsic nature of signals (analog or digital). A general-purpose testing solution does not yet exist off-the-shelf; thus, we developed custom methods and technologies for testing electronic devices. The proposed test methodology has been proven to be both effective and highly flexible, while also simplifying the preparation of experiments. This makes it a powerful tool for evaluating radiation-hardened components under irradiation. Moreover, the same approach and supporting technology can be applied to different types of tests (TID and SEE) as well as to various radiation sources, including gamma rays, protons, and heavy ions. Indeed, although originally designed for TID testing with gamma radiation (from 60Co sources), the methodology has also been successfully employed for SEE testing using both protons and heavy ions.

In this chapter, we focus first on basic concepts of the "two-boards" approach and the proposed design solution by means of an example of memory devices. This chapter continues with some significant examples of test boards/fixtures for the characterization of analog components (ADC, DAC, voltage regulators) developed by the authors. Measurements on analog devices pose more complex problems than digital counterparts, because of the nature of quantities to be measured and the required accuracy. The test setup includes both the hardware ("two-boards") and software (GUI-Graphic User Interface), remote control, and low-level layer) developed custom by the authors. In this section are also presented the Gamma Irradiation Facility build-up at the University of Palermo, which provides the necessary environment for TID testing, and the one at RADEF – University of Jyvaskyla for SEE testing.

5.2 A New Approach of Testing under Radiations

In most cases, the electronic devices under test (DUT) are specifically designed to be radiation-hardened, whereas the associated control electronics may be commercial devices or only radiation-tolerant. This creates a risk of damage to the test control circuitry and introduces uncertainty into the test results, since it can be difficult to distinguish whether an error originates from the device under evaluation or from the test system itself. For this reason, it is crucial to develop irradiation methodologies and build test facilities capable of reproducing the various radiation environments, operational conditions, and signal characteristics (both analog and digital) to which electronic devices may be exposed. At present, no ready-made, general-purpose solution exists; therefore, tailored methods and dedicated technologies have been devised for the testing of electronic components (digital and analog) [6]-[18]. Many of them rely on performing the test within a few hours (from 1 to 3) after the component has been exposed to a radiation dose. This approach does not fully capture the instantaneous reaction of the component, as would be the case with a real-time test. What the authors propose is a real-time measurement of the component's fundamental performance, accepting some limitations in accuracy. But not only that, what is described below is the approach regarding the whole environment (hardware, software, facility, and working method) in which the component testing is inserted in.

To make this concept clearer, it is necessary to start with a typical facility used for the TID test by the authors. On the other side, the test for SEE and SEL with heavy ions poses less critical issues since the ion beam is highly collimated and does not interfere significantly with the DUT control electronics.

5.2.1 The Gamma Irradiation Facility

The Gamma Irradiation Facility of the University of Palermo is a research facility that replicates, on a small scale, the installations of industrial irradiation and, standing midway between the two types, provides a good versatility to perform different types of irradiation. The overall irradiation equipment consists of an irradiation cell, the gamma irradiator IGS-3, and an external control room. The IGS-3 gamma irradiator (see Figure 5.1) is equipped with a total 4,31 TBq of 60Co gamma-ray source.

It is composed of 12 cylindrical shaped sources featuring different activity: six sources of 0,13 TBq each, three sources of 0,24 TBq each, and three sources of 1,26 TBq each. The facility is fully equipped with the instrumentation for dosimetry in order to ensure safe operation and irradiation levels accordingly to international standards ESCC No. 22900 [19].

Inside the cell, we can distinguish two areas: an area of irradiation (realized through direct exposure to radioactive sources), which we might call the "hot room" or irradiation chamber, and a shielded area from radioactive sources (but always within the cell), which we might call "warm room" or access area. In the hot room, radiation fields are very high, and the security systems do not allow anyone to access during irradiation. It is also inadvisable placing any electronic equipment or other devices if not properly shielded, because after a short time they would be severely damaged. In the warm room, the radiation fields are much weaker, and it is possible place

Figure 5.1 Photo of IGS-3 gamma irradiator.

equipment and control instrumentations. The control panel is thus placed outside the irradiation cell, in a fully shielded room (which we might call the "control room"), which allows a "public" and safe access. By considering these stringent conditions of the environment, a new testing strategy shall be defined. The traditional measurements performed using a test board where the device under test is mounted and a lab setup with equipment suitable to generate waveforms and to collect the outputs are not viable. Only the DUT shall be exposed, while the measurement apparatus and auxiliary active circuitry cannot be placed under or near the irradiation beams. A remote control of all measurements operation is mandatory.

5.2.2 The test methodology development

The presence of multiple working areas, each with different access restrictions and radiation levels, complicates the setup of irradiation tests for electronic devices operating in active mode. Power must be supplied through dedicated systems, and the device performance must be monitored using appropriate instruments. Ideally, the control equipment should be in the warm room to prevent potential failures of control electronics from compromising the experiment or producing misleading results. However, this arrangement is not always possible since often the power, auxiliary circuitry, and the monitor functions must be a complementary part of the DUT and shall be placed nearby it. Unfortunately, the degradation of the components driving the DUT becomes noticeable from about 5 krad(Si). For microcontrollers and microprocessors, the main issue is single event functional interrupt (SEFI), which forces frequent reboots of the controller. Level-shifters may introduce incorrect voltage translations, resulting in faulty addressing and control signals, while passive components such as capacitors, resistors, and inductors can exhibit voltage drifts. Beyond 5 krad(Si), active devices, including the FPGA, typically fail, making it impossible to continue the experiment. Since the DUT must withstand up to 300 krad(Si), or even more, it is essential to implement new strategies. In such cases, suitable shielding must be implemented. This serves both to reduce the radiation dose in sensitive areas and to enable proper connection of the equipment to the outside. Shielding involves not only protecting the equipment itself but also arranging the cabling in such a way as to avoid "radiation leaks" into the shielded zone. At the same time, the experiment setup must be designed for full remote operation from the control room, without requiring personnel access to the irradiation cell – since any intervention would necessitate the withdrawal of the sources and the interruption of the test. Finally, the entire setup must comply with ESCC Basic Specification No. 22900 [19] and MIL-STD-883H Method 1019.8 [20].

This objective was achieved by ensuring accurate placement of the DUT and by progressively establishing the necessary connections through the different areas (the hot room, the warm room, and the control room). The adopted solution relies on dividing the test board into two parts:

- a mother-board (MB), which houses the signal generator, data acquisition system, and auxiliary circuitry;
- a daughterboard (DB), which carries only the DUT.

The DB is linked to the MB through flat cables for digital signals, with the MB hosting all control electronics. The flat cable represents a critical element, as in standard testing it is generally avoided due to the potential communication delays it introduces between the MB and DB. However, this limitation is less problematic in the present case, since electronic components designed for space applications typically operate at relatively low frequencies. This reduced speed stems from design strategies adopted to ensure radiation tolerance (or radiation hardness), which inherently restrict the maximum operating frequency. So that a critical element such as a 52-pole flat cable can be still acceptable, provided that its length does not exceed a maximum of half a meter and can run without the significant degradation frequency in the range of 50 MHz. Over this length, a different cable must be taken into account, or the test should be carried out at lower frequency (e.g., 25 MHz). For analog signals, such as those fed as the input to data converters, coaxial cables and SMA connectors shall be adopted, but, of course, a reduced number of pins can be driven.

Both the MB and DB must be carefully designed. In its most basic configuration, the MB incorporates a field-programmable gate array (FPGA, Xilinx Spartan-3®), which generates all the waveforms required by the DUT, handles output data acquisition, manages system control, and interfaces with the laptop via USB (Figure 5.2). Alternative and cheaper solutions foresee the use of microcontroller (i.e., Microchip®) or Arduino®. The latter has been widely used recently, as we will see later. The setup with FPGA is well suited, for instance, to the all-digital testing of radiation-hardened memories [21]-[23].

More advanced versions of the MB also integrate power-generation circuits, enabling it to deliver the necessary supply voltages to the DUT via flat cable, thereby eliminating the need for external power sources. When required by the test, the MB can be further expanded with more sophisticated auxiliary circuitry to generate reference analog signals and provide all the input/output channels needed for evaluating analog rad-hard components such as data converters and power MOSFETs. In these cases, the two boards are interconnected through additional flat and coaxial cables.

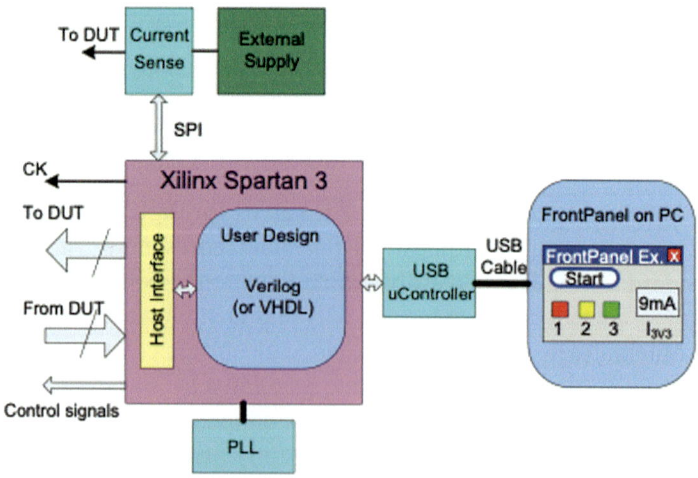

Figure 5.2 Mother-board simplified block diagram.

Although the DB has a simpler role, its design must also be precise. It must not include active devices (e.g., microcontrollers, microprocessors, level shifters) or sensitive passive elements (such as electrolytic capacitors, resistors, or inductors), since their degradation under radiation could compromise the accuracy of the test results. Both board layouts must be optimized, with respect to the ground and the power supply planes, to guarantee the most stable possible working boundary conditions.

As an example, in Figure 5.3 are shown the two boards and main connections for the test of a 1 Mbit SRAM. This component is a low-voltage, high-performance, asynchronous rad-hard 128 kbit × 8 memory device, using six transistor (6T) memory cell with Miller capacitors [21]. It has been integrated in a 180-nm standard CMOS process with a dedicated rad-hard-by-design (RHBD) approach using edge-less transistors (ELT) and enhanced guard-rings. The DB communicates via flat cable with the MB, but all the power supplies are independent. So, in this case, it is necessary to provide a stabilized external power supply that does not bring any noise to the DUT. This latter is hosted in a zero-insertion force (ZIF) socket to give mechanical stability to the entire structure. The whole daughter-board can be considered passive with the DUT as the unique active element, which will be the only target of the incident radiation.

In Figure 5.4, the precise positioning of the above-described experimental setup within the "hot room" of the Gamma Facility is shown. It can be noted that the distance of the DUT from the gamma source is a key point and that a dosimeter is placed nearby the DUT to evaluate with accuracy the delivered dose rate (left). The MB is shielded by means of lead bricks (right).

Figure 5.3 Photo of MB (right) and DB (left) and relevant connection.

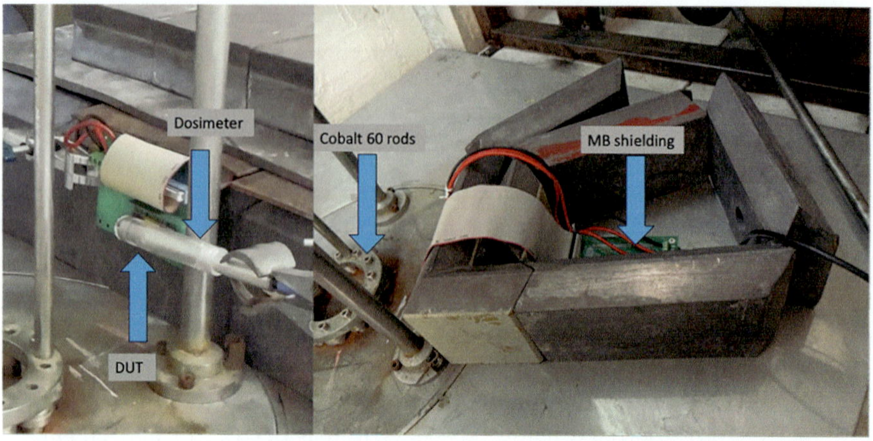

Figure 5.4 Placement in the "hot room" of MB (protected) and DB (exposed) at the Gamma Ray IGS-3 facility (Palermo, Italy).

However, although MB is the most sensible element to be protected during the irradiation, we must consider that it needs to communicate with a personal computer to allow the operator to control the experiment. The low dose rate on the DB (typically 1.2 rad/s(Si)) and the radiation attenuation at a value of about one hundred times lower on the shielded MB ensures the survival of the on-board USB interface needed to communicate with a personal computer to allow the operator to take control of the experiment. The connections with the equipment inside the "hot chamber" can be extended (by using long cables), and once started, the experiment cannot be modified without stopping it. The connections between the "hot room" and the "warm one" are obtained via cable (USB connection). In the "warm room," dose rate is almost zero and electronic equipment is not damaged, but still personnel can't stay. This is why operators are in a "control room" and the connection between it and the "warm room" is obtained via coaxial cable or Ethernet cable.

Figure 5.5 summarizes the methodological approach of the above-described technological test environment.

5.2.3 Remote control testing

The activation of the operations on the MB is organized in two parts: the remote control link and the control software tool. All the equipment within the "warm room" is remotely managed from the "control room" (or "control room"). The remote control of the entire test session was carried out realizing a virtual private network (VPN), able to connect the control room by means

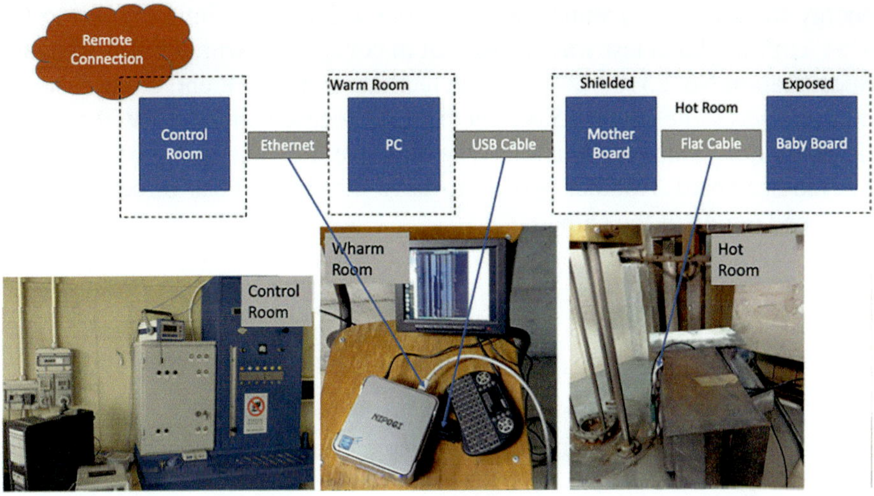

Figure 5.5 The logical split of environments and hardware connections.

of an appropriate commercial software (VNC server®, TeamViewer®, etc.) enabling an external network connection (password protected) in the following steps:

- a connection takes place from the external network to the "cold room";
- a second connection is established from the "cold room" to the PC active within the "warm room";
- the PC accesses via USB link to the mother-board placed inside the "hot room."

A basic VPN configuration was carried out and successfully tested with multiple connections from Milan (the authors in RedCat Devices) and from other locations to the control room of the Palermo facility, where IGS-3 irradiator is located.

The testing setup illustrated in Figure 5.4 provides a high degree of flexibility and simplifies the preparation of experiments. Its adaptability and effectiveness have been validated in multiple irradiation facilities with varying layouts, radiation sources, and test types (both TID and SEE), all while applying the same methodology. A series of irradiation campaigns were conducted by the authors with heavy ions at the Radiation Effects Facility (RADEF), University of Jyväskylä, Finland, and with protons at the Svedberg Laboratory (TSL), Uppsala University, Sweden. At RADEF, different ions were employed to investigate both soft and hard errors, with linear energy transfer (LET) values ranging from 4–5 MeV·cm²/mg (Si) up to 60 MeV·cm²/mg. Each test was carried out in compliance with the ESCC Basic Specification No. 25100 [24]. Source facilities were different, but test methodology was same (Figure 5.6). The beam is focused on the die without any lid protection. Since the beam is very directive, it is not necessary to shield the mother-board. Another goal is to have the Bragg Peak directly on the front-end region of the chip.

5.2.4 Software tools

As concerns software tools, we have seen that for the complete environment, we need:

- a commercial software for the remote control and for establishing a VPN;
- an interface on the PC in the "warm room" to make it easy for the operator to control the experiment;
- a low-level SW to program the FPGA, the microcontroller, or the Arduino®.

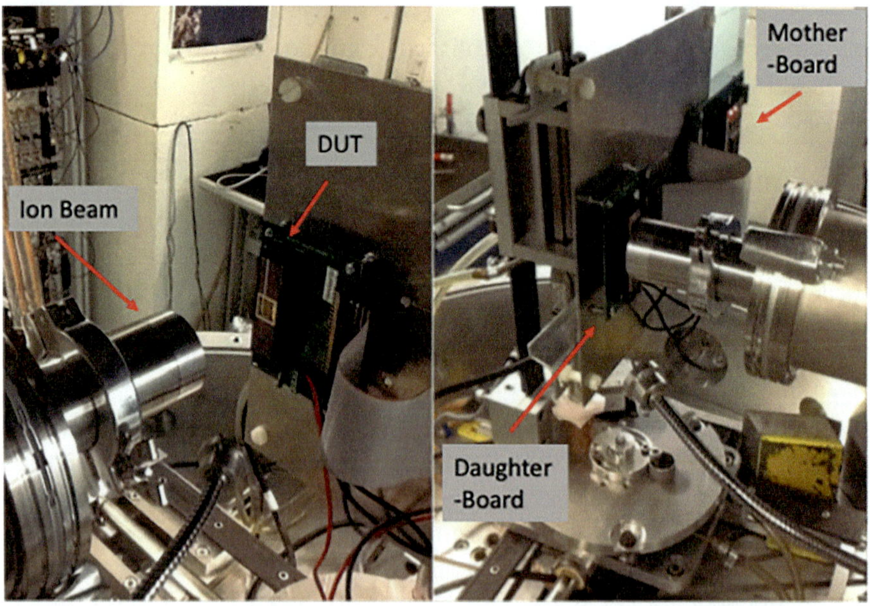

Figure 5.6 The testing setup in a heavy ions irradiation facility in Jyvaskyla.

At the time of the authors' first tests, when simpler solutions were not yet available, it was necessary to use an FPGA as a controller, so that a Verilog code has been written to generate all functions that shall be provided by FPGA. Moreover, a suitable GUI has been developed in Visual Basic to control all operations from the PC through USB. It is based on a proprietary interface (Opal Kelly's FrontPanel) that handles all interaction between the virtual controls and the FPGA. Figure 5.7 shows how the memory test tool looks like. The memory test allowed us to apply a very different pattern such as the checkerboard, the negate checkerboard, the March, and other more complex tests and evaluate the number of bit failure coming from the degradation of the memory array. This operation is carried out during the irradiation and controlled by the automatic procedures (software routines) or directly by the operators connected to the board.

However, when simpler control devices, such as Arduino®, have been available on the market, the low-level software and the GUI have been evolved toward cheaper, more flexible solutions, as will be discussed in the next three paragraphs:

- Test of rad-hard ADC;
- Test of rad-hard DAC;
- Test of rad-hard BGRs.

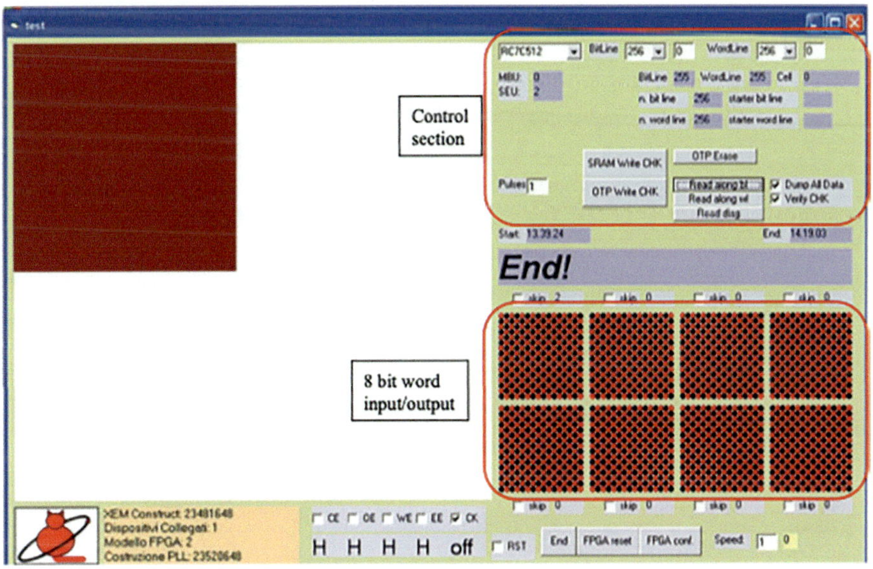

Figure 5.7 Screenshot of the control tool for memory testing.

5.3 Test of a 3.3V 130-nm CMOS Rad-hard 12-Bit ADC

The first example of testing under radiation is the radiation-hardened 12-bit ADC developed in a 130-nm CMOS standard process by the authors and described in Chapter 4. The component layout with padring is shown in Figure 5.8. The ADC has 12-bit output (B0 to B11), one control signal (Start-PowerDown), and the Clock. The EoC (End-of-Conversion) signal indicates that the ADC output bits are stable.

It was tested under Cobalt-60 irradiation at the University of Palermo facility, using the two-board concept and RedCat Devices proprietary Leonida Platform, which includes the overall software environment. At a dose rate of 0.63 rad/s, the ADC withstood up to 350 krad(Si) without showing any significant performance degradation.

As said, the proposed solution involves splitting the test board into two parts:

- a mother-board, which hosts the signal generation, data acquisition system, and auxiliary circuitry;
- a daughter-board, carrying only the DUT.

The DB connects to the MB – where the control electronics are located – via a flat cable for digital signals and coaxial cables for critical analog signals. Being an "analog measurement," the approach is rather more complicated

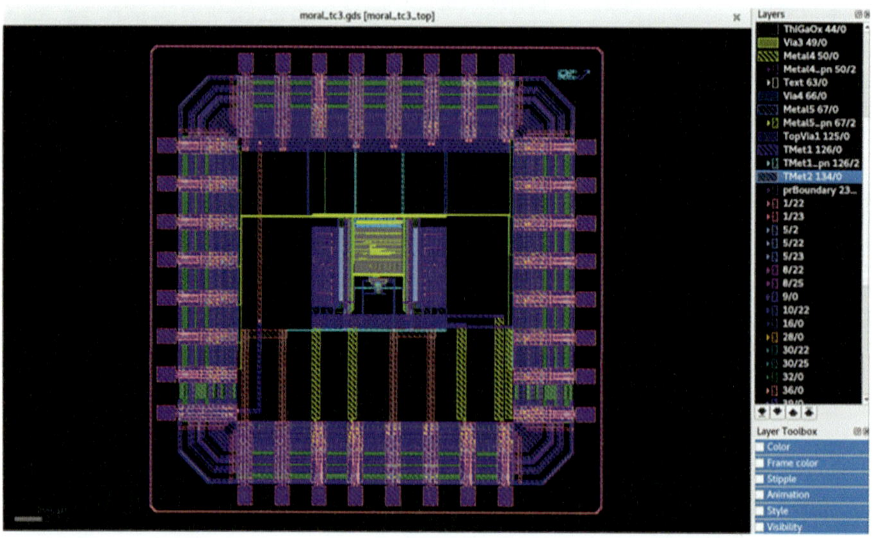

Figure 5.8 Layout with padring of the 12-bit ADC.

with respect to that for memories. Different design approaches are possible for the MB. It can be custom-made, offering maximum design freedom but at higher cost and with the need to develop dedicated control software. Alternatively, an off-the-shelf board can be used, which is less flexible but more cost-effective and benefits from existing development tools. In this work, we adopted an off-the-shelf solution, employing an Arduino Due® (Arduino S.r.l., Monza, Italy) [25]. To extend its functionality, we designed a small shield board (to be mounted directly on the Arduino) with additional features (Figure 5.9).

The Arduino Due® provides on-board two 12-bit programmable DACs for generating the analog differential signals for the ADC under test. Then five ADCs (ADC0 to ADC4) monitor the reference voltages (VRP and VRN), common mode voltage (VMC), and the differential input signals fed to the ADC (INP and INN). Finally, ADC5 and ADC6 monitor the analog and digital current consumption of the DUT. Arduino Due® generates Clock and Start signals as well. The 12-bit coming from the designed ADC are captured and stored in a file. The shield board complements this setup with two low-noise buffers for DUT differential input signals, three programmable resistive dividers (for generating VRP, VRN, and VCM), and current sense, along with LDO regulators for power supply (3.3V) and current sensing. The latter is essentially made up of a simple resistor, R_{sense}, across whose terminals the voltages are measured. Dividing the voltage difference by R_{sense}, we get the

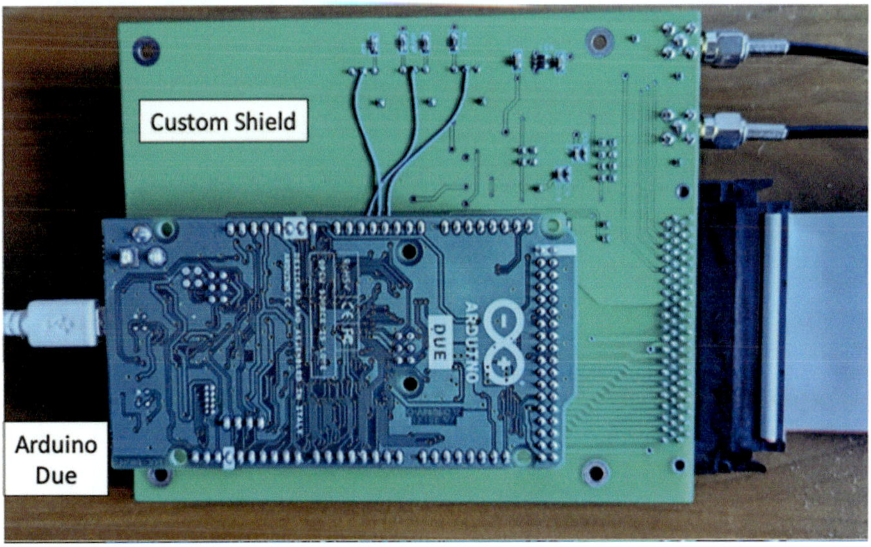

Figure 5.9 Detail of the Arduino 2 board plus the custom shield.

current consumption. We have two sense resistors, one for analog and the other for digital.

A detailed block diagram of the three boards employed for testing under radiation is shown in Figure 5.10, while a photo of the complete system is presented in Figure 5.11.

As shown in this setup, a flat cable is used to handle the digital signals, while two coaxial cables connect the DACs' analog differential input to the Arduino's ADC. Because the operating frequencies are relatively low, a detailed signal integrity analysis is not required, except in specific situations such as the clock of a data converter.

The test facility itself is arranged into three separate areas, as described in the previous paragraph: a "hot room," a "warm room," and a "control room." The "hot room" houses the IGS-3 gamma irradiator (Figure 5.1), which contains a 60Co gamma-ray source, along with the two test boards. The three rooms are connected with USB and Ethernet connections, and it is also possible to remotely control the PCs in the control chamber by means of a remote terminal application.

Inside the "hot chamber," the Arduino Due® mother-board is protected using lead brick shielding, while the daughter-board containing the DUT

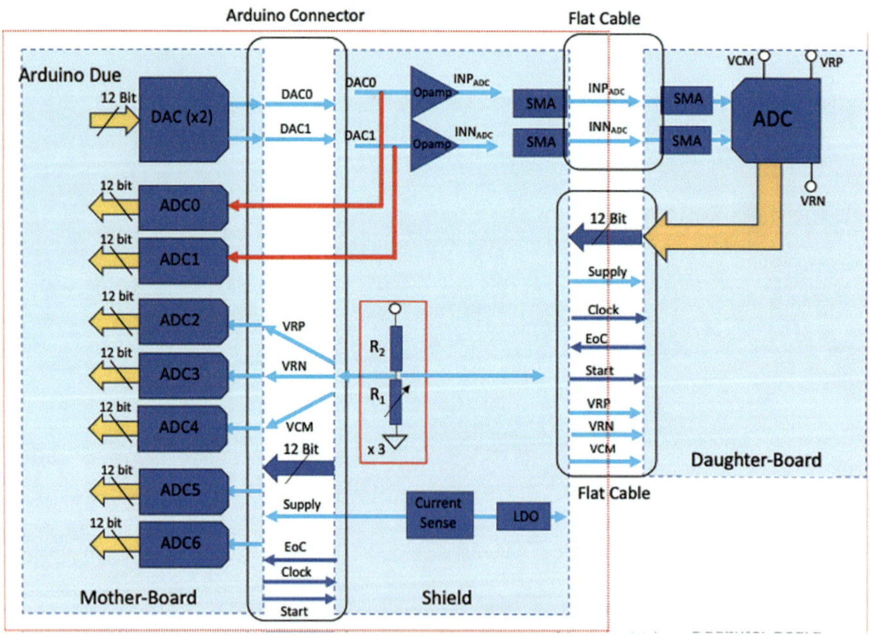

Figure 5.10 Block diagram of the boards used for ADC tests under irradiation.

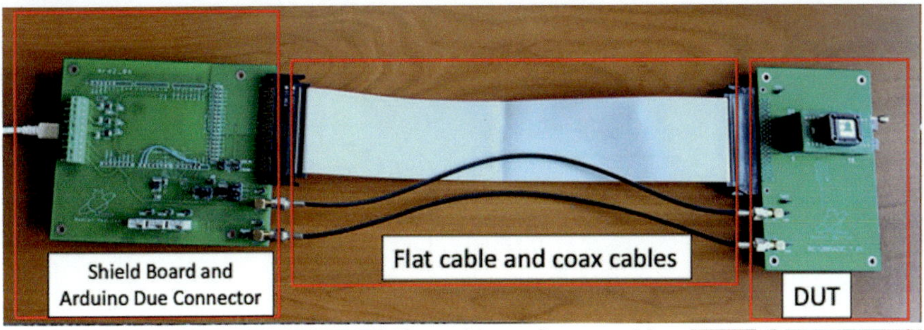

Figure 5.11 Photo of the complete system (MB + DB) for the TID test of the ADC.

remains directly exposed. Figure 5.12 shows the final arrangement, where the MB is secured within the closed shielding box and the DB is left exposed. A dosimeter positioned near the DUT (not visible in the photo) is used to verify the exposure rate. The shielding enclosure (assembled from lead bricks) in the back protects the MB placed inside.

The approach used for measurements in this specific case is the one shown in Figure 5.13. The test bits are applied to the DAC hosted on Arduino; the DAC output analog voltage is fed to the DUT; the output bits of the designed

Figure 5.12 Placement of the MB (protected) and DB with ADC (exposed) with the shielding box for MB (with lead bricks) in the background.

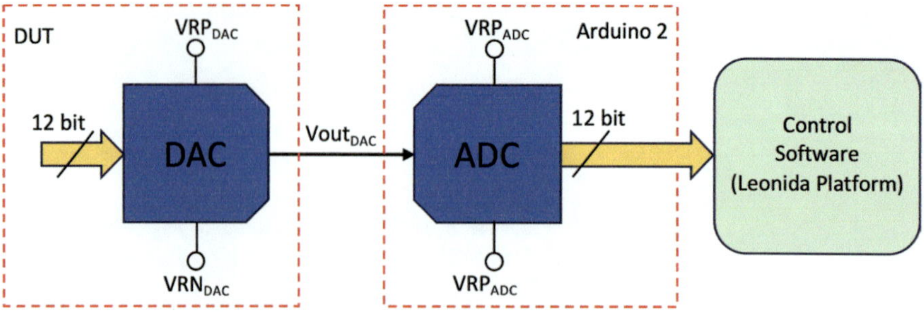

Figure 5.13 Test block diagram for measuring the ADC digital output.

ADC are recorded remotely by the developed software (Leonida Platform Ver. 20230301) and converted again in analog by means of the following equation:

$$(VRP_{ADC} - VRN_{ADC}/4095) \times ADC_{outcode} + V_{offset,ADC} \tag{5.1}$$

where VRP_{ADC} and VRN_{ADC} are ADC references, $ADC_{outcode}$ is the measured digital code, and $V_{offset,ADC}$ is the offset of the ADC.

The Leonida Platform software is organized in two layers. The first one is low-level, based on IDE environment by Arduino®, which generates command directly fed to Arduino by an USB connection. Arduino firmware is written in C using the IDE interface available with the board, and, thanks to the internal microcontroller, many logic signals (3.3 or 5 V according to the board release) can be programmed together with internal ADCs (12) and DACs (2), when needed. A picture of the IDE environment is shown in Figure 5.14.

Arduino IDE has a terminal interface establishing the serial connection between the PC and the Arduino board; in this case, commands are written

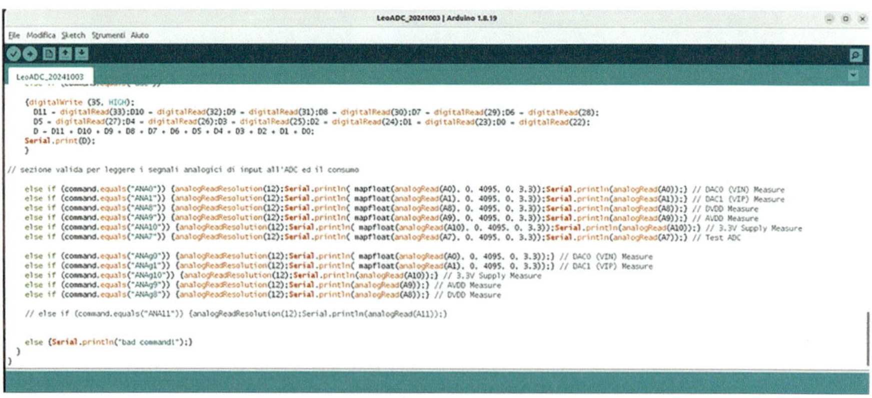

Figure 5.14 Snapshot of the IDE environment: program.

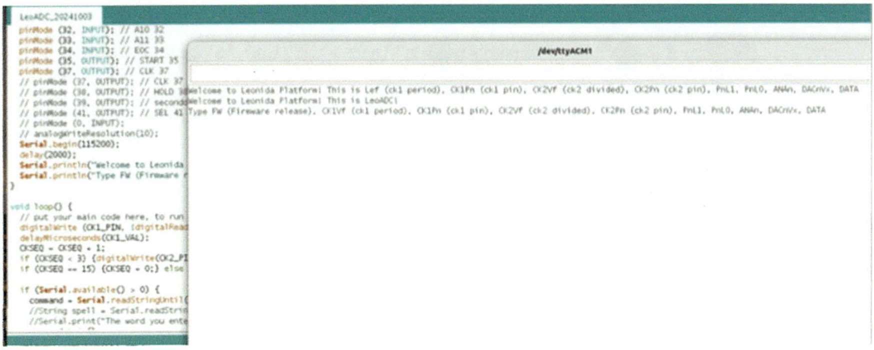

Figure 5.15 Snapshot of the IDE environment: terminal interface.

and sent directly to the USB interface. For more complex applications, a software using firmware-based commands are recommended to make the testing procedure easier (Figure 5.15).

The second SW layer is arranged as a graphical user interface, which allows users to control the test and monitor measurements results. This Leonida Platform code is capable not only of generating individual codes but also of producing code ramps and sinewaves, which are converted by the two Arduino DACs and fed to the designed ADC (the waveforms are shown in Figure 5.16(a)–(e)).

The figures below show the platform's GUI, which enables remote control of multiple aspects of the experiment. For instance, the top section allows users to define the filename for storing acquired data (the ADC output bit), select the type of input signal to be applied to the ADC (sinusoid, ramp, etc.), and specify the number of samples. The plots in the center display the chosen input waveform (DACs output in blue, ADC input in red, ADC output after conversion in analog in green).

On the left-hand side of the top graph, users can set the positive, negative, and common mode voltage references for the ADC. Below that, it is possible to set the positive and negative input range of the differential input signals. Then, the interface also provides the option to apply a fixed DC value to the ADC and immediately monitor the corresponding output bit. Then, power consumption can be monitored. Finally, in the bottom-right side, the dose rate and cumulated irradiation dose from the beginning of the experiment are displayed.

We don't report here the obtained test results, since the details are present in Chapter 4. We would only remind that using a 0.63 rad/s dose rate, during the experiment the ADC reached up to 350 krad(Si) with no significant degradation in terms of electrical performance. Any failure has been

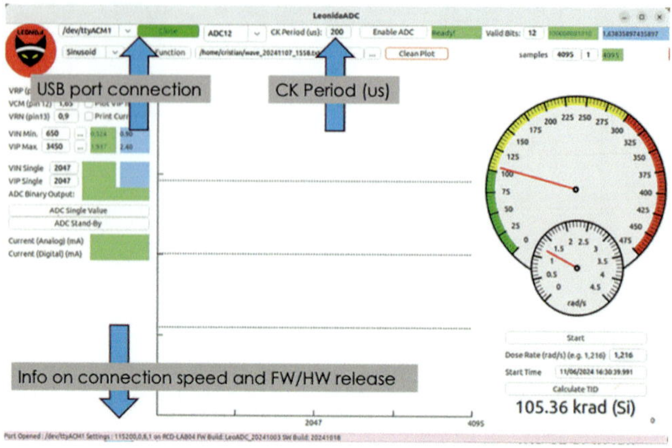
LeonidaADC
USB port connection
CK Period (us)
Info on connection speed and FW/HW release
105.36 krad (Si)

(a)

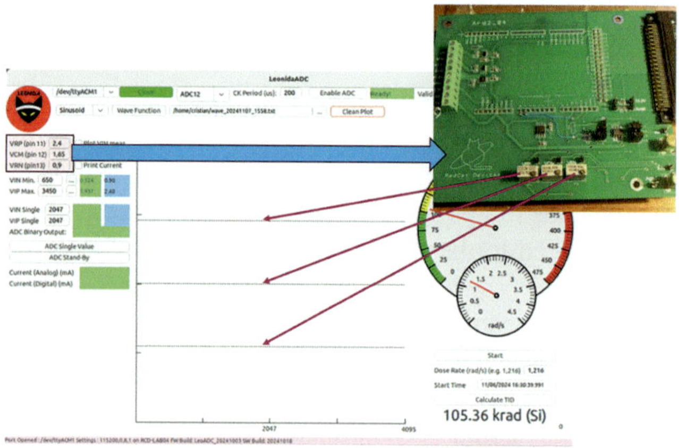
LeonidaADC
105.36 krad (Si)

(b)

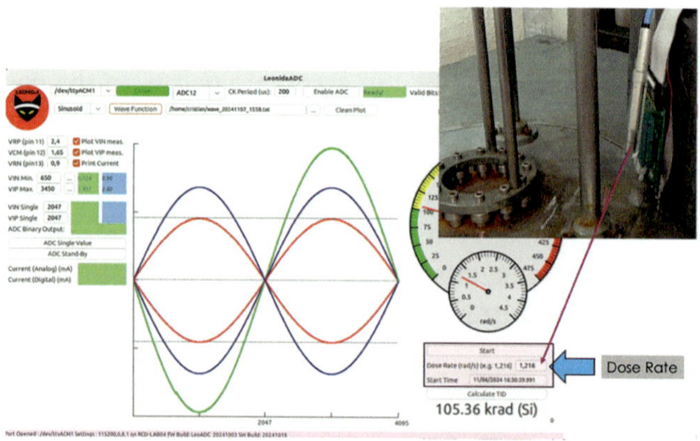
LeonidaADC
Dose Rate
105.36 krad (Si)

(c)

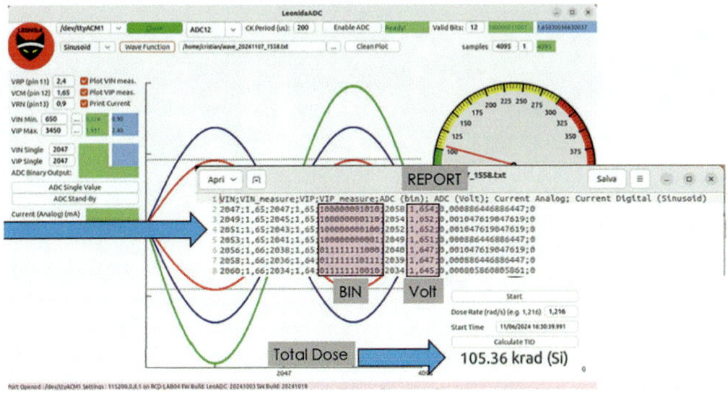

(d)

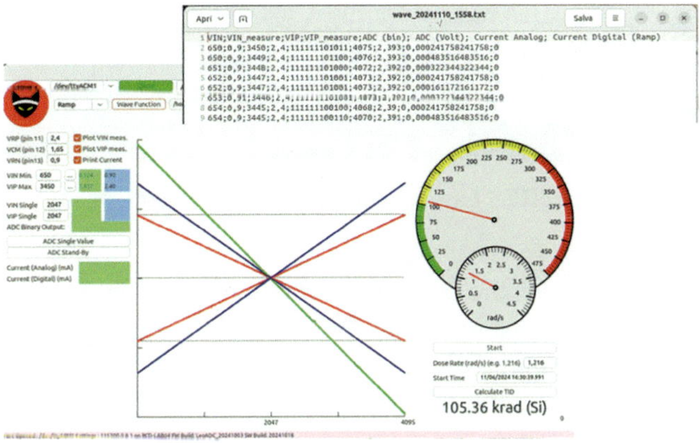

(e)

Figure 5.16 (a)–(e) Snapshot of the Leonida Platform for ADC.

observed in the test setup (mother-board was fully functional at the end of the experiment) with the adopted approach.

5.4 Test of a 3.3V 130-nm CMOS Rad-hard 12-Bit DAC

The second example of testing under radiation is the radiation-hardened 12-bit DAC developed in a 130-nm CMOS standard process by the authors and described in Chapter 3. The component layout with padring is shown in Figure 5.17. The DAC has a 12-bit output (A0 to A11), three control signals (Reset, Hold, Power-Down), the Clock, and one Select signal that allows to by-pass the output buffer if needed. The analog signal DOUT is the DAC analog output voltage.

It was tested under Cobalt-60 irradiation at the University of Palermo facility, using the two-board concept and RedCat Devices proprietary Leonida

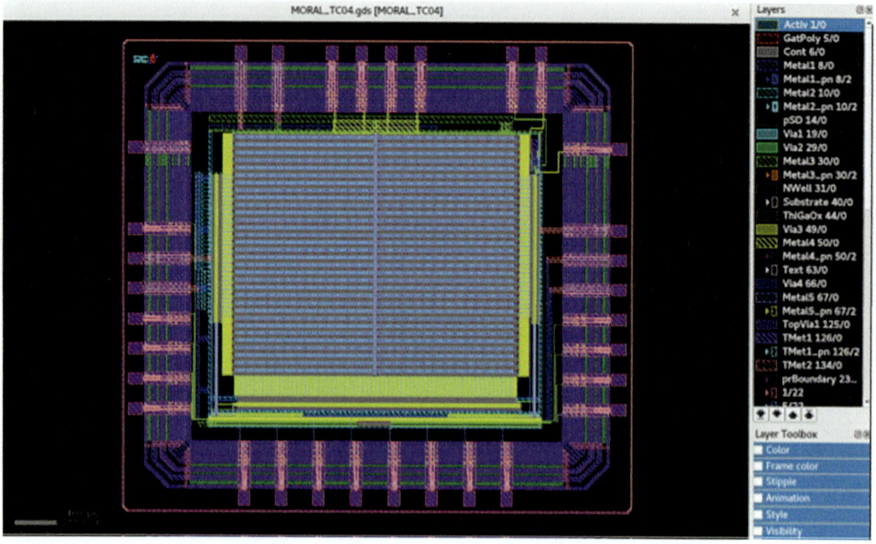

Figure 5.17 Layout with padring of the 12-bit DAC.

Platform, which includes the overall software environment. At a dose rate of 0.63 rad/s, the DAC withstood up to 500 krad(Si) without showing any significant performance degradation.

As said, the proposed solution involves splitting the test board into two parts:

- a mother-board, which hosts the signal generation, data acquisition system, and auxiliary circuitry;
- a daughter-board, carrying only the DUT.

The DB connects to the MB – where the control electronics are located – via a flat cable for digital signals and coaxial cables for critical analog signals. Also in this work, we adopted an Arduino Due® (Arduino S.r.l., Monza, Italy). Also in this case, we designed a small shield board (to be mounted directly on the Arduino) with additional features (Figure 5.18). The Arduino Due provides on-board digital signal generators (for bit generation), precise ADCs for monitoring the 12-bit DAC's output voltage and current consumption, and two programmable DACs to supply voltage references to the DUT. The shield complements this setup with two low-noise buffers for DUT references and current measurement, along with LDO regulators for power supply (3.3V) and current sensing. The latter is made up with a simple resistor, R_{sense}, connected in series to the power supply, across

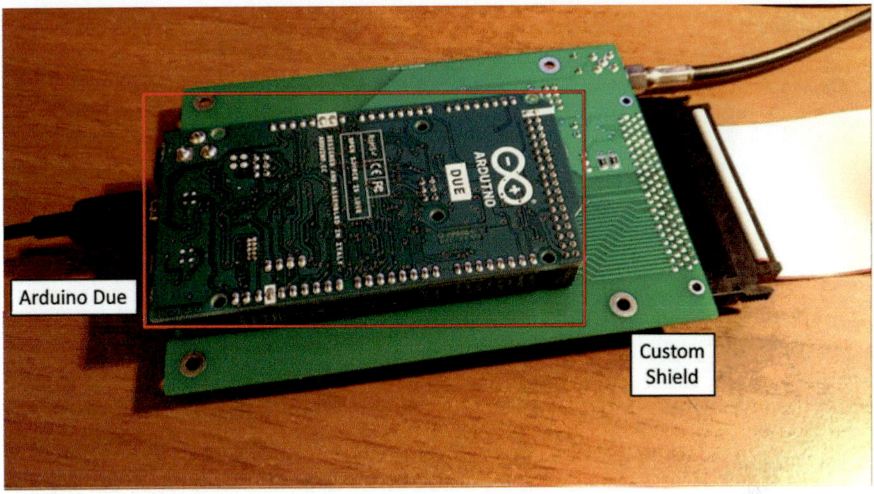

(a)

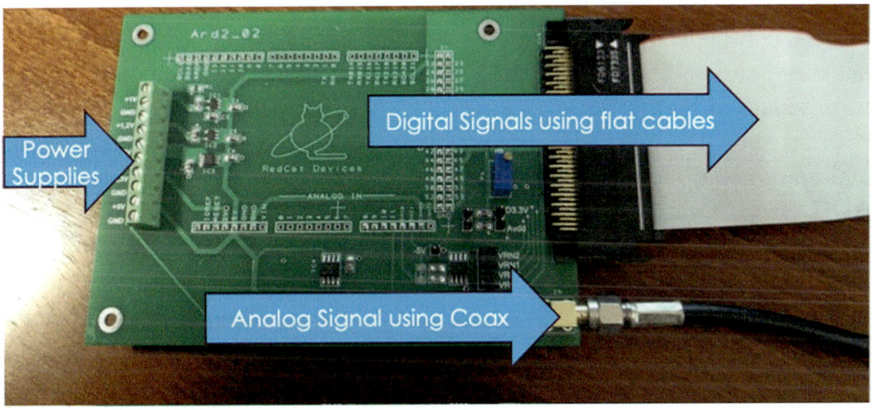

(b)

Figure 5.18 Detail of the Arduino Due board plus the custom shield [27]: (a) recto; (b) verso.

whose terminals the voltages are measured. Dividing the voltage difference by R_{sense}, we get the current consumption. We have two sense resistors, one for analog and the other for digital supply. Finally, the shield hosts also a potentiometer for the regulation of the bias current of the DAC output buffer. The shield is directly connected with Arduino boards and represents the intermediate connection between the microcontroller and the DUT hosted on the daughter-board.

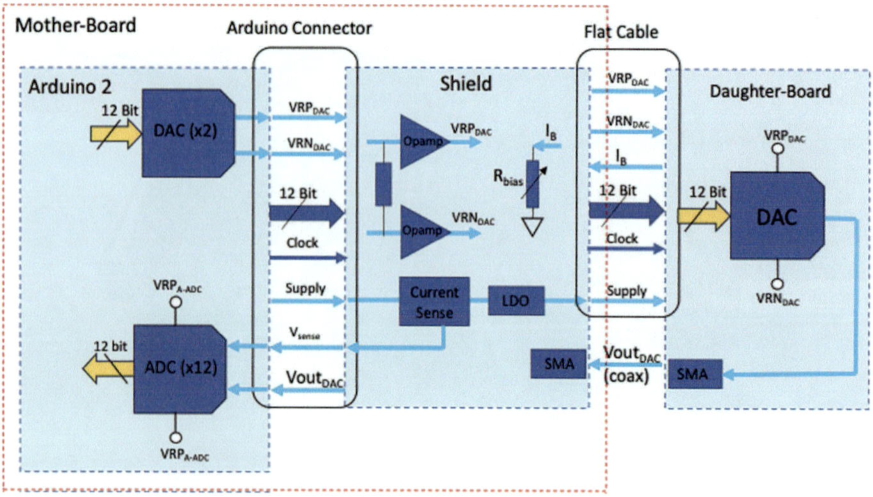

Figure 5.19 Block diagram of the boards used for DAC tests under irradiation [27].

A complete block diagram of the three boards employed for testing under radiation is shown in Figure 5.19, while a photo of the complete system is presented in Figure 5.20.

As shown in this setup, a flat cable is used to handle the digital signals, while a coaxial cable connects the DAC's analog output to the Arduino's ADC. Because the operating frequencies are relatively low, a detailed signal integrity analysis is not required, except in specific situations such as the clock of a data converter.

The test facility itself was arranged into three separate areas, as described in the previous paragraph: a "hot room," a "warm room," and a "control room." The "hot room" houses the IGS-3 gamma irradiator (Figure 5.1),

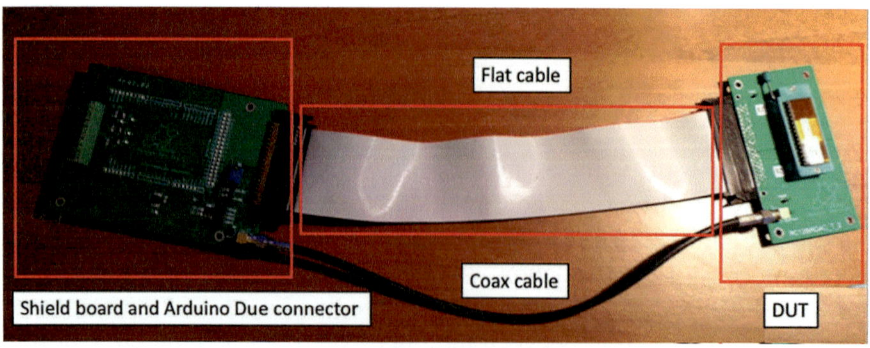

Figure 5.20 Photo of the complete system (MB + DB) for the TID test of the DAC [27].

which contains a 60Co gamma-ray source, along with the two test boards. The three rooms are connected with USB and Ethernet connections, and it is also possible to remotely control the PCs in the control chamber by means of a remote terminal application.

Inside the hot chamber, the Arduino Due® mother-board is protected using lead brick shielding, while the daughter-board containing the DUT remains directly exposed. Figure 5.21(a) illustrates the shielding enclosure (assembled from lead bricks) in its open state, with the MB placed inside (the additional board shown is unrelated to the DAC test). Figure 5.21(b) shows the final arrangement, where the MB is secured within the closed shielding box and the DB is left exposed. A dosimeter positioned near the DUT (not visible in the photo) is used to verify the exposure rate.

The approach used for the measurements of the analog output in this specific case is shown in Figure 5.22. The test bits are applied to the DUT; the DAC output voltage is converted to digital by the ADC hosted in the Arduino Due board, and its output bits are recorded remotely by the developed software (Leonida Platform Ver. 20230301). The obtained data can be kept as they are or converted again in analog by means of eqn (5.1).

The Arduino ADC references (as measured before the test) are fixed and are equal to $VRP_{ADC} = 3.3$ V and $VRN_{ADC} = 30$ mV, and its LSB = 0.798 mV.

We control the experiment with the Leonida Platform software: low-level in IDE environment and high-level with a custom GUI. For the DAC, the Leonida Platform is capable of generating individual codes, and also of producing code ramps, square waves, and sinewaves, as illustrated in Figure 5.23. The figure shows the platform's GUI, which enables remote

(a) (b)

Figure 5.21 Shielding box (with lead bricks) still open with the MB (a); placement of the MB (protected) and DB with DAC (exposed) (b) [27].

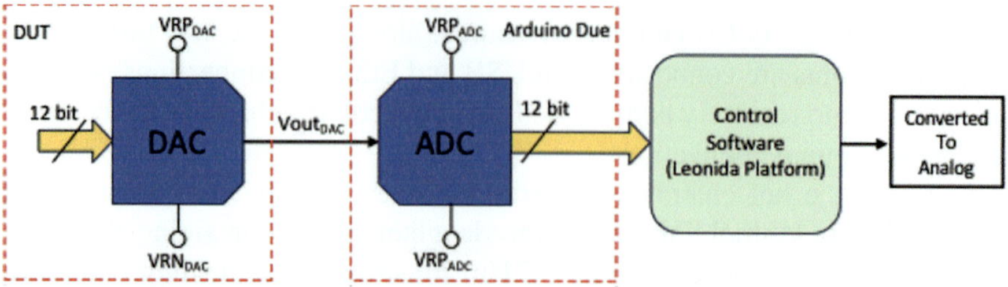

Figure 5.22 Block diagram of the test system for measuring the DAC analog output voltage.

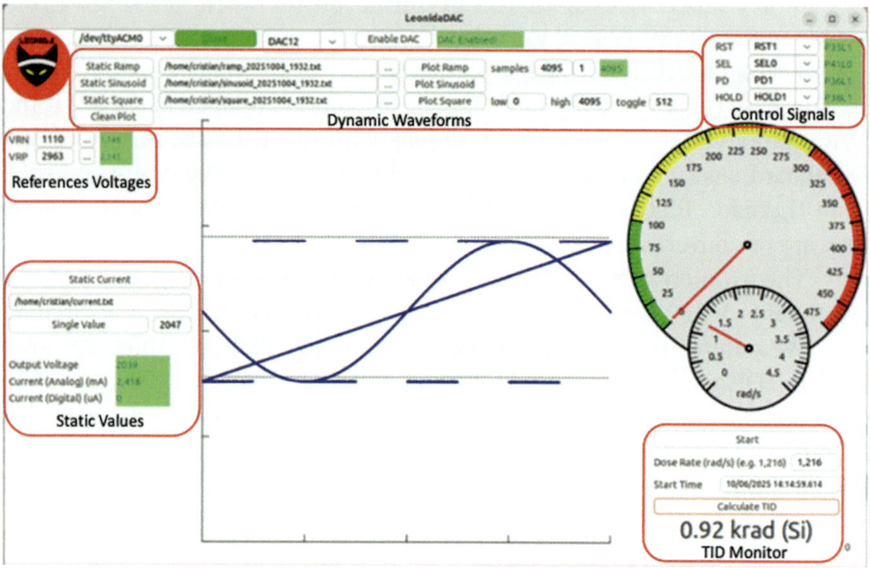

Figure 5.23 Snapshot of the Leonida Platform for DAC.

control of multiple aspects of the experiment. For instance, the top section allows users to define the filename for storing acquired data (the DAC output voltage), select the type of input waveform to be applied to the DAC (ramp, sinusoid, or square wave), and specify the number of samples. The measured DAC output voltages are listed in the storing file, which can be easily imported in a spreadsheet tool and elaborated. The plot commands present in the center display the chosen input waveform for the experiment (all three waveforms are displayed in the figure).

On the left-hand side of the top graph, users can set the positive and negative voltage references for the DAC. Below that, the interface also provides the option to apply a fixed DC value to the DAC and immediately monitor the corresponding output voltage. In the upper right side, it is possible to set the three control signals (RST, HOLD, and PD) and the SEL (Select). Finally, in the bottom right side, the dose rate and cumulated irradiation dose from the beginning of the experiment are displayed.

We do not report here the obtained test results, since the details are present in Chapter 3. We would only remind that using a 0.63 rad/s dose rate, during the experiment the DAC reached up to 500 krad(Si) with no significant degradation in terms of electrical performance. Any failure has been observed in the test setup (mother-board was fully functional at the end of the experiment) with the adopted approach.

5.5 Test of a 3.3-V 180-nm CMOS Rad-hard Bandgap

The third example of testing under radiation is the radiation-hardened BGRs developed by the authors in a 180-nm CMOS standard process and described in Chapter 2. The bandgap references are present in a test chip that includes also two rad-hard LDOs, not relevant for our discussion. The component layout with padring is shown in Figure 5.24, where only the circuits circled in red are relevant.

Three bandgaps are fixed, while the other three, besides supply (which is also the input) and ground, have four Trim<0:3> digital signals to perform

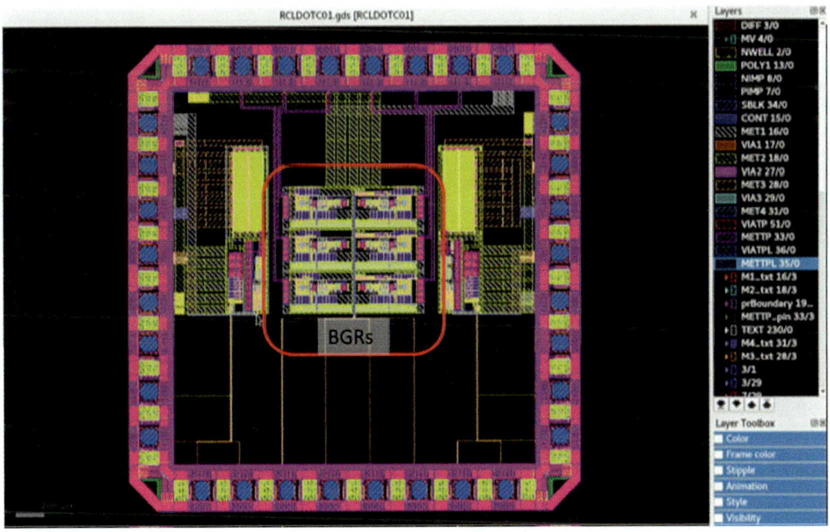

Figure 5.24 Layout with padring of the test chip including the developed BGRs.

the fine regulation of their output voltages. Output voltages are indicated with BGR_XXF, where XX is the generated voltage and F means that the output is filtered, as already explained in Chapter 2.

The test chip was tested under Cobalt-60 irradiation at the University of Palermo facility, using the two-board concept and RedCat Devices proprietary Leonida Platform, which includes the overall software environment. Using a 0.75 rad/s dose rate, during the experiment, the BGRs reached up to 700 krad(Si) with no significant degradation in terms of electrical performance.

Once again, the proposed solution involves splitting the test board into two parts:

- a mother-board, which hosts the signal generation, data acquisition system, and auxiliary circuitry;
- a daughter-board, carrying only the test chip.

The DB connects to the MB – where the control electronics are located – via a flat cable for digital signals and static analog signals. The daughter-board with the socket for the DUT is shown in Figure 5.25(a). Also in this work, as the mother-board we adopted an off-the-shelf solution, employing an Arduino Due® (Arduino S.r.l., Monza, Italy). To extend its functionality, we designed a small shield board (to be mounted directly on the Arduino) with additional features (Figure 5.25(b)).

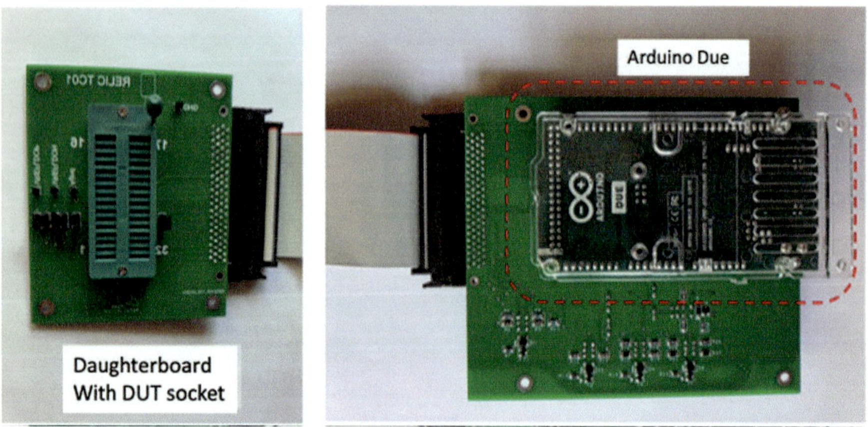

(a) (b)

Figure 5.25 Detail of the daughter-board with DUT socket (a) Arduino Due board plus the custom shield (b).

The Arduino Due® provides on-board 12 ADCs for the conversion of analog outputs coming from the DUT (six for BGRs) and two DACs for auxiliary voltage generations (for LDOs, not relevant in this book). It provides also the needed supplies. The last two ADCs read the currents sensing output (analog and digital), implemented with resistors like in the two previous examples. Arduino Due® generates the Trim digital signals (for BGRs), as well. Besides the current sensing circuits, the shield hosts some auxiliary circuits for the LDOs.

A detailed block diagram of the three boards employed for testing under radiation is shown in Figure 5.26, while a photo of the complete system is presented in Figure 5.27.

As shown in this setup, a flat cable is used to handle the digital signals and, in this case, also the analog signals. Indeed, coaxial cables are not needed because these signals are essentially static.

For TID, the test facility itself is arranged into three separate areas, as described for the DAC test, and the overall environment is the same, as well (Figure 5.28).

We do not report here the obtained test results, since the details are present in Chapter 2. We would only remind that using a 0.75 rad/s dose rate, during the experiment, the BGRs reached up to 700 krad(Si) with no significant degradation in terms of electrical performance (output voltage and power

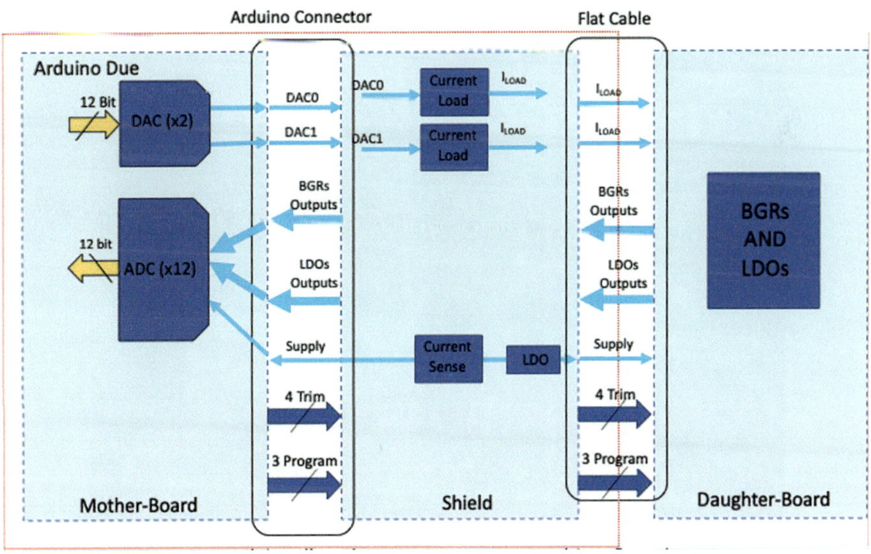

Figure 5.26 Block diagram of the boards used for BGRs tests under irradiation.

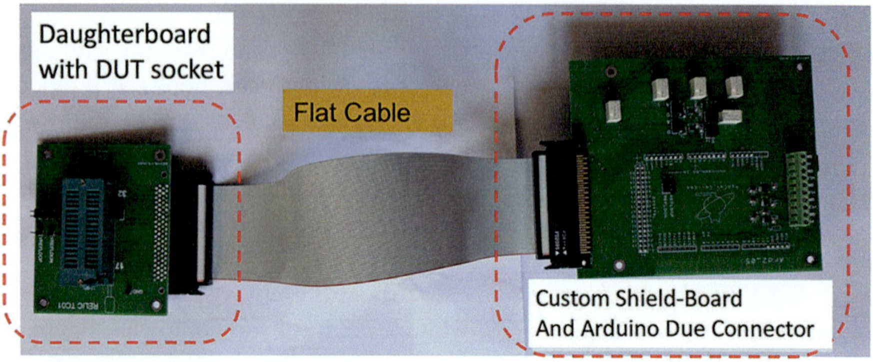

Figure 5.27 Photo of the complete system (MB + DB) for the TID test of the BGRs.

consumption). Any failure has been observed in the test setup (mother-board was fully functional at the end of the experiment) with the adopted approach.

For the BGRs, the authors performed also the test of the BGRs with heavy ion at RADEF University of Jyvaskyla (Finland). The relevant setup is shown in Figure 5.29. The characteristics of the beam were the following:

- Ions: Xenon
- Energy: 2059 MeV

Figure 5.28 Placement of the MB (protected) and DB with BGRs (exposed) with the shielding box for MB (with lead bricks) in the background.

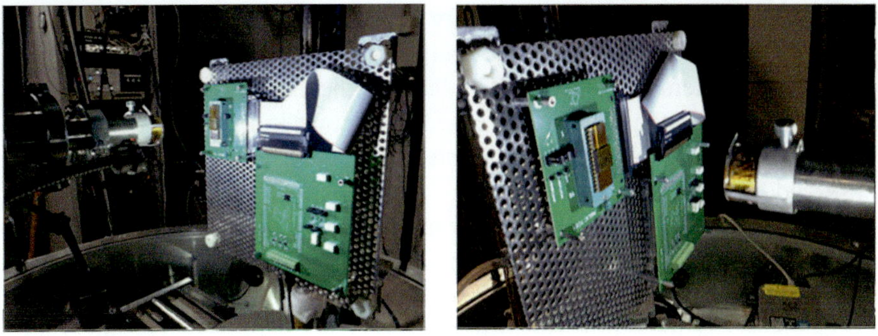

(a) (b)

Figure 5.29 Test setup for heavy ion measurements: (a) overall view; (b) detailed view with the DUT.

- Flux: 900 #ions/s
- LET: 60 MeV*cm^2/mg

As said, since the beam is very collimated, it was not needed to shield the mother-board as in the TID test. As can be seen in Figure 5.30, the experiment

Figure 5.30 Complete environment for heavy ions test: beam room and control room (Jyvaskyla-FI), remote control (Milan-IT).

has been controlled both in the facility control room and remotely. The former is needed to check that the Leonida software works correctly and to monitor the parameters of the beam. The latter is the Leonida Platform software for the control of the test. In this case, the picture in the lower right corner shows the control room at Milan premises. Any SEL on the test chip has been observed during the experiment.

Also, for the BGRs test, the Leonida Platform software is organized in two layers, the lower based on IDE environment by Arduino, which generates command directly fed to Arduino by a USB connection, and the second one in the language like the Basic. The former is shown in Figure 5.31. The uppermost is the terminal command window, while the lowest is the program window.

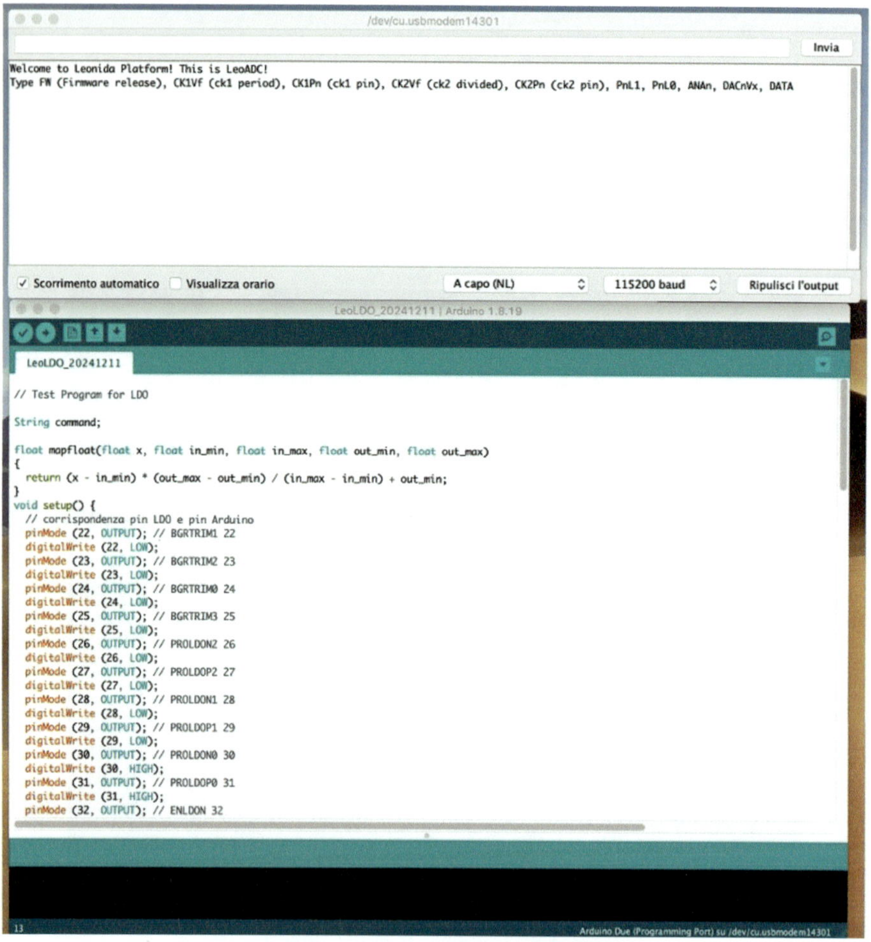

Figure 5.31 Snapshot of low-level SW developed in the IDE environment by Arduino.

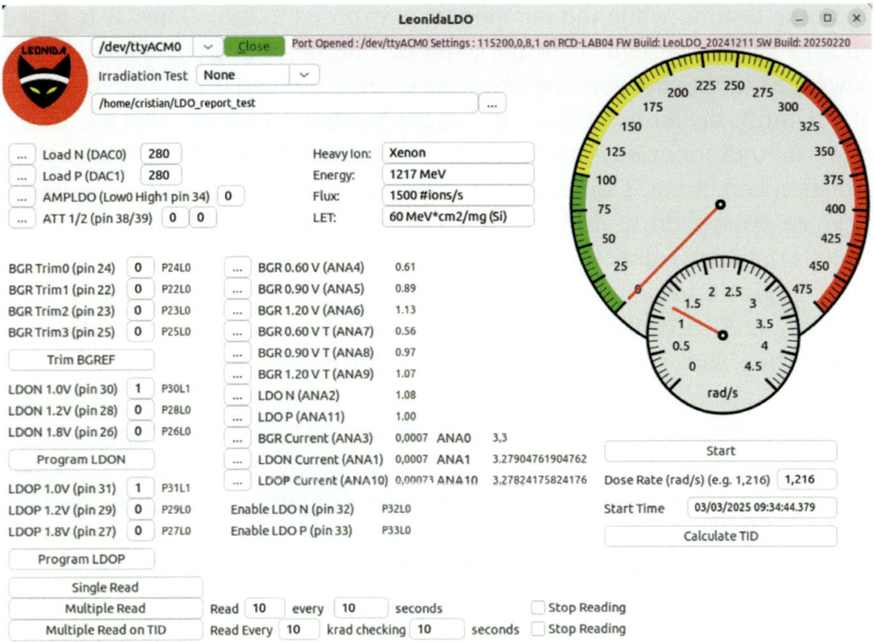

Figure 5.32 Snapshot of the Leonida Platform for BGRs and LDOs.

The Leonida Platform GUI is arranged as a graphical user interface, which allows users to control the test and monitor measurements results. It can support both the test of BGRs and LDOs (not relevant for our discussion) and is shown in Figure 5.32. In the GUI, some functions can be identified: in red for setting, in blue for measurement results, and in green for control. The section N. 1 surrounded in red (1–4) allows to set the value of the current for an active load (100 uA, 10 mA, or 20 mA) useful for LDOs, and N. 2, 3 and 4 drives the bit for the trimming of bandgap output and programming the LDOs output voltage. Section 5 in blue presents the measured BGRs output voltages, the LDOs output voltage, and their current consumptions. Finally, Section 6 in green controls the measure, as single shot or multiple read once every given krad. Other features in the right-bottom side include dose rate and TID, while on the top-center, it is possible to define the file name where measured data are stored.

5.6 Conclusion

In this chapter, we have analyzed the basic concept of testing of rad-hard component under total dose and single-events (i.e., heavy ions). Unlike the widely used methods up to now, the authors propose a novel approach for real-time testing, while the component is exposed to radiations. It is based on a two-board system, with the exposed one hosting the DUT only. The developed testing methodology has proven effective and offers a high degree of flexibility, thanks to the ease of preparing test experiments. This makes it a powerful tool for evaluating radiation-hardened components under available radiation conditions. Consequently, the same testing approach and technology can be applied to different types of tests (such as TID and SEE) and various radiation sources, including gamma rays, protons, and heavy ions.

Originally designed for TID tests using gamma radiation (from ^{60}Co sources) on SRAM, the methodology has also been successfully adapted for SEE testing with protons and heavy ions. Several test boards and software tools have already been developed and validated for a variety of components. This methodology is demonstrated by means of three examples, a 12-bit DAC, a 12-bit ADC and BGRs, which have been illustrated in this chapter. However, owing to the flexibility of the proposed layout, the testing methodology can be extended to other component classes, such as microprocessors, and phase-locked loops (PLLs), among others.

The authors consider particularly promising the possibility, enabled by this approach, of performing tests under very low dose-rate conditions. Such tests typically require long durations (sometimes lasting several months) and can be conducted not only with ^{60}Co sources but also with alternative sources such as ^{241}Am. Very low dose-rate testing is especially suitable for replicating space conditions on Earth, and when combined with a fully automated system featuring periodic remote monitoring, it offers a cost-effective solution for the qualification of space components.

It is worth emphasizing that the main advantage of this methodology lies in its in-situ approach, which provides researchers with a powerful tool for performing various component-level tests during irradiation sessions. This results in greater flexibility and reduced session costs, since components do not need to be removed from the radiation chamber and transferred to a separate, shielded testing environment. As a final remark, the remotization of the experiment control offers another advantage in terms of flexibility, avoiding travel costs.

References

[1] Mavis, D. G., Al ex an der D. R., Employing Radiation Hardness by Design Techniques with Commercial te grated Circuit Processes, Digital Avionics Systems Conference, (1997), Oct., pp. 15-22

[2] Clark, L., et al., "Optimizing Radiation Hard by Design SRAM Cells", IEEE Trans. on Nucl. Sc., 54 (2007), 6, pp. 2028-2036

[3] L. Clark, et al., "Optimizing Radiation Hard by Design SRAM Cells", IEEE Trans. on Nucl. Sc., vol. 54, N. 6, Dec. 2007, pp. 2028 – 2036

[4] N. F. Haddad, at al., "Incremental enhancement to SEU hardened 90nm CMOS memory cell", Proc. of the 11th Europ. Conf. on Radiation and its Effects on Components and Systems (RADECS 2010), Sep. 2010.

[5] Haddad, N. F., et al., "Incremental Enhancement to SEU Hardened 90 nm CMOS Memory Cell", Proceedings, 11th Europ. Conf. on Radition and its Effects on Components and Systems (RADECS 2010), Aqua Dome Langenfeld, Austria, Sep. 20-24, 2010

[6] L. Bonora, "Overview of a Space Qualification Process - Radiation characterization and beyond", ESA/ESTEC-Component Division, Xhanti (Greece), Oct. 2006

[7] K. Kruckmeyer, "Challenges of Single Event Upset and Transient Testing and Characterization of Mixed Signal Products", Overview of National Semiconductor's radiation programs, May 18, 2012

[8] R.L. Pease, et al., "Radiation Testing of Semiconductor Devices for Space Electronics", December 1988, Proceedings of the IEEE Vol. 76, pp. 1510 - 1526

[9] R. Bannatyne, "Radiation testing of CMOS devices required for space use", Military Embedded Systems, Sept. 11, 2017

[10] C. B. Polo, "ASIC Radiation Testing Methodology and Techniques", 4th International Workshop on Analogue and Mixed-Signal Integrated Circuits for Space Applications AMICSA 2012, Aug. 26 – 28, ESTEC - Noordwijk - The Netherlands

[11] A. Parlato, et al., "The methodology for active testing of electronic devices under the radiations", Nuclear Technology and Radiation Protection Vol. 33(1), March 2018, pp. 53-60

[12] D. Chen, et al., "Total-Ionizing Dose and Single-Event Effect Test Results of Analog and Mixed Signal Devices from Analog Devices", 2018 IEEE Radiation Effects Data Workshop (REDW), 16-20 July 2018, Waikoloa, HI, USA

[13] Véronique Ferlet-Cavrois, "Essentials for electronics in the space radiation environment - Test methods and facilities", 18 May 2021, ESA Internal Course, EEE Component Radiation. Hardness Assurance Tutorial, ESTEC - Noordwijk - The Netherlands

[14] M. N. Al-Haddad, at al., "Development and Testing of a Flash Analog-to-Digital Converter Based System for Pulse Shape Discrimination of

Nuclear Radiation Pulses", IEEE Trans. on Nucl. Science, Vol. 41, N. 5, Oct. 1994, pp. 1765-1769

[15] V. Agarwal, et al., "Comparison of gamma radiation performance of a range of CMOS A/D converters under biased conditions", IEEE Transactions on Nuclear Science, Vol.: 52, N. 6, Dec. 2005, pp. 3059-3067

[16] K. Kruckmeyer, et al., "Use of Code Error and Beat Frequency Test Method to Identify Single Event Upset Sensitive Circuits in a 1 GHz Analog to Digital Converter", IEEE Transactions on Nuclear Science, Vol. 55, N. 4, Aug. 2008, pp. 2013-2018

[17] S. Danzeca, et al., "New Testing Methodology of an Analog to Digital Converter for the LHC Mixed Radiation Field", IEEE Trans. on Nucl. Science, Vol. 60, N. 4, Aug. 2013, pp. 2598-2604

[18] A.B. Karakozov, et al., "Single Event Effects and Total Dose Testing of Digital to Analog Converters", 2017 17th European Conference on Radiation and Its Effects on Components and Systems (RADECS)

[19] Total Dose Steady-State Irradiation Test Method, ESCC Basic Specfication n. 22900, Is sue 4, 2010

[20] MIL-STD-883H, Method 1019.8, Ionizing Radiation (To tal Dose) Test Procedure, 2010.

[21] Calligaro, C., et al., "A 15 Mrad(Si) 512 Kbit Rad-hard SRAM in a Standard 0.18 mm CMOS Technology, Proceedings", 13th Europ. Conf. on Radiation and its Effects on Components and Systems (RADECS 2012), Biarritz, France, Sep. 24-28, 2012

[22] Arbat, A., et al., "Radiation Hardened 2Mbit SRAM in 180 nm CMOS Technology", IEEE-AESS Conference in Europe about Space and Satellite Telecommunication 2012 (ESTEL), Nov. 2012

[23] Calligaro, C., et al., A Multi-Megarad, Radiation Hardened by Design 512 kbit SRAM in CMOS Technology, 22nd International Conference on Microeletronics (ICM2010), Il Cairo (Egypt), Dec., 2010, pp. 375-378

[24] Single Event Effect test Method and Guidelines, ESCC Basic Specification No. 25100, Issue 2, 2014

[25] Available online: hXps://docs.arduino.cc/hardware/due/ (accessed on 18 April 2023)

Index

About the Authors

Dr. Umberto Gatti was born in Pavia (Italy) in 1962. He received Laurea and Ph.D. degrees in electronics engineering from the University of Pavia in 1987 and 1992, respectively. From 1993 to 1999, he worked in the R&D Lab of Italtel, as ASIC Designer, and was involved in modeling analog ICs. In 1999, he joined the Development Technologies Lab of Siemens, where he was Sr. Design ASIC Engineer. Besides developing telecom ICs, he was the coordinator of Eureka-Medea+ projects focused on high-speed sigma-delta converters. In 2008 he moved to Nokia Siemens Networks where he worked as Sr. Power Supply Architect. Since 2012 he has been with RedCat Devices as a member of the executive staff. During his career he has been involved in the design of data-converters, broadband wireless transceivers, Hall sensors micro-systems and power supply architectures. His current research interests are in rad-hard CMOS ICs, particularly rad-hard libraries and memories (SRAMs and OTP), rad-hard mixed-signal circuits (ADC-DAC), rad-hard linear voltage regulators, and in testing such components. He holds 2 international patents and is co-author of about 60 papers and one book (Rad-hard Semiconductor Memories, River Publishers). He is Senior Member of IEEE and serves also as reviewer for magazines and conferences.

Dr. Cristiano Calligaro received his laurea degree in electronics engineering and his Ph.D. degree in electronics and information technology engineering from the University of Pavia (Italy) in 1992 and 1997 respectively. In 2006 he established RedCat Devices srl as a start-up. During his career he has been involved in memory design (volatile and non-volatile) both for consumer application (multilevel flash memories) and space applications (rad-hard memories) and software design for SEE/TID evaluation using open source EDA tools. His current research interest is focused on rad-hard standard cell libraries to be used for rad-hard mixed signal ASICs, stand-alone memory compilers and testing methodologies for rad-hard components. He holds more than 20 patents mainly in the field of multilevel NVMs and he is co-author of more than 60 papers and one book (Rad-hard Semiconductor Memories, River Publishers). He has been coordinator of RAMSES and

ATENA projects inside the Italy–Israel Cooperation Programme, SkyFlash project in the European FP7 Programme, EuroSRAM4Space project in the Eureka Eurostars Programme and RADPROM project funded by Italian Space Agency (ASI). He is an IEEE Senior Member and an IEEE Transactions on Nuclear Science reviewer.

For Product Safety Concerns and Information please contact our
EU representative GPSR@taylorandfrancis.com Taylor & Francis
Verlag GmbH, Kaufingerstraße 24, 80331 München, Germany